—WORKBOOK—

Applied Math

FOR WATER
PLANT OPERATORS

JOANNE KIRKPATRICK PRICE
Training Consultant

CRC Press
Taylor & Francis Group
Boca Raton London New York

CRC Press is an imprint of the
Taylor & Francis Group, an **informa** business

Cover art adapted from photograph of Fallbrook Sanitary District Reclaimed Water System.

CRC Press
Taylor & Francis Group
6000 Broken Sound Parkway NW, Suite 300
Boca Raton, FL 33487-2742

© 1991 by Taylor & Francis Group, LLC
CRC Press is an imprint of Taylor & Francis Group, an Informa business

No claim to original U.S. Government works

ISBN-13: 978-0-87762-875-0 (pbk)

Visit the Taylor & Francis Web site at
http://www.taylorandfrancis.com

and the CRC Press Web site at
http://www.crcpress.com

Library of Congress Cataloging-in-Publication Data

Main entry under title:
  Workbook—Applied Math for Water Plant Operators

Full Catalog record is available from the Library of Congress

Dedication

This book is dedicated to my family:

To my husband Benton C. Price who was patient and supportive during the two years it took to write these texts, and who not only had to carry extra responsibilities at home during this time, but also, as a sanitary engineer, provided frequent technical critique and suggestions.

To our children Lisa, Derek, Kimberly, and Corinne, who so many times had to pitch in while I was busy writing, and who frequently had to wait for my attention.

To my mother who has always been so encouraging and who helped in so many ways throughout the writing process.

To my father, who passed away since the writing of the first edition, but who, I know, would have had just as instrumental a role in these books.

To the other members of my family, who have had to put up with this and many other projects, but who maintain a sense of humor about it.

Thank you for your love in allowing me to do something that was important to me.

J.K.P.

Contents

Contents—Cont'd

Contents—Cont'd

Contents—Cont'd

Preface to the Second Edition

The first edition of these texts was written at the conclusion of three and a half years of instruction at Orange Coast College, Costa Mesa, California, for two different water and wastewater technology courses. The fundamental philosophy that governed the writing of these texts was that those who have difficulty in math often do not lack the ability for mathematical calculation, they merely have not learned, or have not been taught, the "language of math." The books, therefore, represent an attempt to bridge the gap between the reasoning processes and the language of math that exists for students who have difficulty in mathematics.

In the years since the first edition, I have continued to consider ways in which the texts could be improved. In this regard, I researched several topics including how people learn (learning styles, etc.), how the brain functions in storing and retrieving information, and the fundamentals of memory systems. Many of the changes incorporated in this second edition are a result of this research.

Two features of this second edition are of particular importance:

- the **skills check section** provided at the beginning of every basic math chapter

- a **grouping of similar types of calculations** in the applied math texts

The skills check feature of the basic math text enables the student to pinpoint the areas of math weakness, and thereby customizes the instruction to the needs of the individual student.

The first six chapters of each applied math text include calculations grouped by type of problem. These chapters have been included so that students could see the common thread in a variety of seemingly different calculations.

The changes incorporated in this second edition were field-tested during a three-year period in which I taught a water and wastewater mathematics course for Palomar Community College, San Marcos, California.

Written comments or suggestions regarding the improvement of any section of these texts or workbooks will be greatly appreciated by the author.

Joanne Kirkpatrick Price

Acknowledgments

"From the original planning of a book to its completion, the continued encouragement and support that the author receives is instrumental to the success of the book." This quote from the acknowledgments page of the first edition of these texts is even more true of the second edition.

First Edition

Those who assisted during the development of the first edition are: Walter S. Johnson and Benton C. Price, who reviewed both texts for content and made valuable suggestions for improvements; Silas Bruce, with whom the author team-taught for two and a half years, and who has a down-to-earth way of presenting wastewater concepts; Mariann Pape, Samuel R. Peterson and Robert B. Moore of Orange Coast College, Costa Mesa, California, and Jim Catania and Wayne Rodgers of the California State Water Resources Control Board, all of whom provided much needed support during the writing of the first edition.

The first edition was typed by Margaret Dionis, who completed the typing task with grace and style. Adele B. Reese, my mother, proofed both books from cover to cover and Robert V. Reese, my father, drew all diagrams (by hand) shown in both books.

Second Edition

The second edition was an even greater undertaking due to many additional calculations and because of the complex layout required. I would first like to acknowledge and thank Laurie Pilz, who did the computer work for all three texts and the two workbooks. Her skill, patience, and most of all perseverance has been instrumental in providing this new format for the texts. Her husband, Herb Pilz, helped in the original format design and he assisted frequently regarding questions of graphics design and computer software.

Those who provided technical review or assistance with various portions of the texts include Benton C. Price, Kenneth D. Kerri, Lynn Marshall, Wyatt Troxel, Mike Hoover, Bruce Grant and Jack Hoffbuhr. Their comments and suggestions are appreciated and have improved the current edition.

Many thanks also to the staff of the Fallbrook Sanitary District, Fallbrook, California, especially Virginia Grossman, Nancy Hector, Joyce Shand, Mike Page, and Weldon Platt for the numerous times questions were directed their way during the writing of these texts.

The staff of Technomic Publishing Company, Inc., also provided much advice and support during the writing of these texts. First, Melvyn Kohudic, President of Technomic Publishing Company, contacted me several times over the last few years, suggesting that the texts be revised. It was his gentle nudging that finally got the revision underway. Joseph Eckenrode helped work out some of the details in the initial stages and was a constant source of encouragement. Jeff Perini was copy editor for the texts. His keen attention to detail has been of great benefit to the final product. Leo Motter had the arduous task of final proof reading.

I wish to thank all my friends, but especially those in our Bible study group (Gene and Judy Rau, Floyd and Juanita Miller, Dick and Althea Birchall, and Mark and Penny Gray) and our neighbors, Herb and Laurie Pilz, who have all had to live with this project as it progressed slowly chapter by chapter, but who remained a source of strength and support when the project sometimes seemed overwhelming.

Lastly, the many students who have been in my classes or seminars over the years have had no small part in the final form these books have taken. The format and content of these texts is in response to their questions, problems, and successes over the years.

To all of these I extend my heartfelt thanks.

How To Use These Books

The *Mathematics for Water and Wastewater Treatment Plant Operators* series includes three texts and two workbooks:

- Basic Math Concepts for Water and Wastewater Plant Operators

- Applied Math for Water Plant Operators

- Workbook—Applied Math for Water Plant Operators

- Applied Math for Wastewater Plant Operators

- Workbook—Applied Math for Wastewater Plant Operators

Basic Math Concepts

All the basic math you will need to become adept in water and wastewater calculations has been included in the Basic Math Concepts text. This section has been expanded considerably from the basic math included in the first edition. For this reason, students are provided with more methods by which they may solve the problems.

Many people have weak areas in their math skills. It is therefore advisable to take the skills test at the beginning of each chapter in the basic math book to pinpoint areas that require review or study. If possible, it is best to resolve these weak areas <u>before</u> beginning either of the applied math texts. However, when this is not possible, the Basic Math Concepts text can be used as a reference resource for the applied math texts. For example, when making a calculation that includes tank volume, you may wish to refer to the basic math section on volumes.

Applied Math Texts and Workbooks

The applied math texts and workbooks are companion volumes. There is one set for water treatment plant operators and another for wastewater treatment plant operators. Each applied math text has two sections:

- Chapters 1 through 6 present various calculations **grouped by type of math problem**. Perhaps 70 percent of all water and wastewater calculations are represented by these six types. Chapter 7 groups various types of pumping problems into a single chapter. The calculations presented in these seven chapters are common to the water and wastewater fields and have therefore been included in both applied math texts.

  Since the calculations described in Chapters 1 through 6 represent the heart of water and wastewater treatment math, if possible, it is advisable that you master these general types of calculations before continuing with other calculations. Once completed, a review of these calculations in subsequent chapters will further strengthen your math skills.

- The remaining chapters in each applied math text include calculations **grouped by unit processes**. The calculations are presented in the order of the flow through a plant. Some of the calculations included in these chapters are not incorporated in Chapters 1 through 7, since they do not fall into any general problem-type grouping. These chapters are particularly suited for use in a classroom or seminar setting, where the math instruction must parallel unit process instruction.

The workbooks support the applied math texts section by section. They have also been vastly expanded in this edition so that the student can build strength in each type of calculation. A detailed answer key has been provided for all problems. The workbook pages have been perforated so that they may be used in a classroom setting as hand-in assignments. The pages have also been hole-punched so that the student may retain the pages in a notebook when they are returned.

The workbooks may be useful in preparing for a certification exam. However, because theses texts include both fundamental and advanced calculations, and because the requirements for each certification level vary somewhat from state to state, it is advisable that you <u>first determine the types of problems to be covered in your exam</u>, then focus on those types of calculations in these texts.

1 *Applied Volume Calculations*

PRACTICE PROBLEMS 1.1: Tank Volume Calculations

1. The diameter of a tank is 80 ft. If the water depth is 30 ft, what is the volume of water in the tank, in gallons?

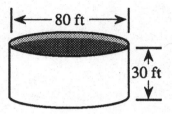

ANS_____

2. The dimensions of a tank are given below. Calculate the cubic feet volume of the tank.

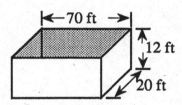

ANS_____

3. A tank 25 ft wide and 80 ft long is filled with water to a depth of 13 ft. What is the volume of water in the tank (in gal)?

ANS_____

4. What is the volume of water in a tank, in gallons, if the tank is 15 ft wide, 30 ft long, and contains water to a depth of 10 ft?

ANS_____

5. Given the tank diameter and depth shown below, calculate the volume of water in the tank, in gallons.

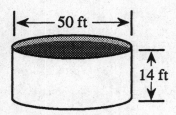

ANS_____

PRACTICE PROBLEMS 1.2: Channel or Pipeline Capacity Calculations

1. What is the cubic feet volume of water in the section of rectangular channel shown below?

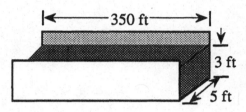

ANS_____

2. A new section of 10-inch diameter pipe is to be disinfected before it is put into service. If the length of pipeline is 1500 ft, how many gallons of water will be needed to fill the pipeline?

ANS_____

3. Calculate the gallon volume of the section of trapezoidal channel shown below.

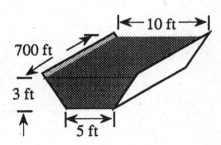

ANS_____

4. A section of 6-inch diameter pipeline is to be filled with chlorinated water for disinfection. If 1778 ft of pipeline is to be disinfected, how many gallons of chlorinated water will be required?

ANS_____

5. What is the volume of water (in gal) for the 1000-ft section of channel shown below?

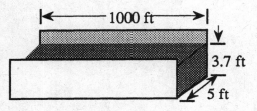

Volume = 138,380 gallons

ANS_____

PRACTICE PROBLEMS 1.3: Other Volume Calculations

1. A trench is to be excavated that is 3 ft wide, 3.5 ft deep, and 600 ft long. What is the cubic yards volume of the trench?

ANS_____

2. A pond is 6 ft deep. Given the dimensions as shown below, calculate the cu ft volume of the pond.

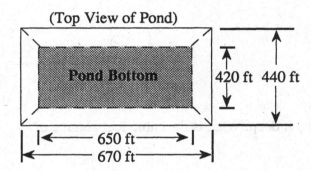

(Top View of Pond)

Pond Bottom

420 ft   440 ft

650 ft
670 ft

ANS_____

3. Given the dimensions of the trench shown below, calculate the cubic yard volume of the trench.

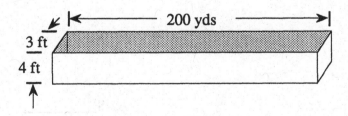

200 yds

3 ft
4 ft

ANS_____

4. Calculate the cu ft volume of the oxidation ditch shown below. The cross section of the ditch is trapezoidal.

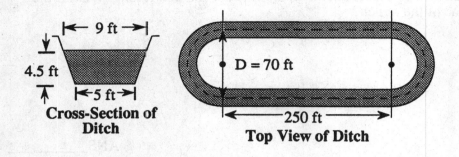

9 ft

4.5 ft

5 ft

Cross-Section of Ditch

D = 70 ft

250 ft

Top View of Ditch

ANS_____

5. A trench is 250 yards long, 2 ft wide and 2 ft deep. What is the cubic feet volume of the trench?

250 yds

2 ft

2 ft

ANS_____

Chapter 1—Achievement Test

1. What is the cubic feet volume of water in the rectangular channel shown below?

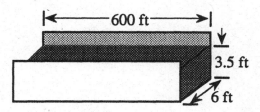

ANS_____

2. The diameter of a tank is 70 ft. If the water depth in the tank is 23 ft, what is the volume of water in the tank, in gallons?

ANS_____

3. A pond is 4 ft deep. Given the dimensions shown below, calculate the cubic feet volume of water in the pond.

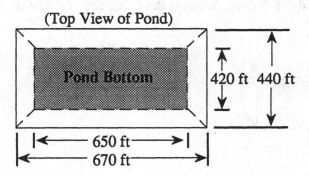

(Top View of Pond)

Pond Bottom

420 ft 440 ft

650 ft
670 ft

ANS_____

4. The dimensions of a tank are given below. Calculate the cubic feet volume of the tank.

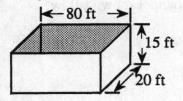

ANS_____

5. A new section of 8-inch diameter pipe is to be filled with water for testing. If the length of pipeline is 3500 ft, how many gallons of water will be needed to fill the pipeline?

ANS_____

6. A trench is 300 yards long, 2 ft wide, and 2.5 ft deep. What is the cubic feet volume of the trench?

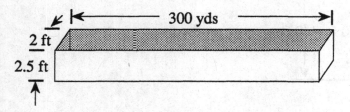

ANS_____

Chapter 1—Achievement Test (Cont'd)

7. A trench is to be excavated. If the trench is 2.5 ft wide, 3 ft deep and 1500 ft long, what is the cubic yards volume of the trench?

ANS_____

8. Calculate the maximum gallon capacity of the section of trapezoidal channel shown below.

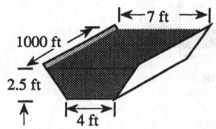

ANS_____

9. A tank is 20 ft wide and 60 ft long. If the tank contains water to a depth of 13 ft, how many gallons of water are in the tank?

ANS_____

10. What is the volume of water (in gal) contained in a 2000-ft section of channel if the channel is 6 ft wide and the water depth is 3.7 ft? (Round the circumference length to the nearest foot.)

ANS_____

11. Calculate the cu ft capacity of the oxidation ditch shown below. The cross-section of the ditch is trapezoidal.

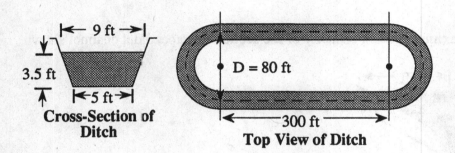

Cross-Section of Ditch

Top View of Ditch

ANS_____

12. Given the diameter and water depth shown below, calculate the volume of water in the tank, in gallons.

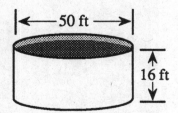

ANS_____

2 *Flow and Velocity Calculations*

PRACTICE PROBLEMS 2.1: Instantaneous Flow Rates

1. A channel 42 inches wide has water flowing to a depth of 2.6 ft. If the velocity of the water is 2.2 fps, what is the cfm flow in channel?

ANS_____

2. A tank is 15 ft long and 10 ft wide. With the discharge valve closed, the influent to the tank causes the water level to rise 0.7 feet in one minute. What is the gpm flow to the tank?

ANS_____

3. A trapezoidal channel is 3.5 ft wide at the bottom and 5.5 ft wide at the water surface. The water depth is 38 inches. If the flow velocity through the channel is 125 ft/min, what is the cfm flow rate through the channel?

ANS_____

4. A 6-inch diameter pipeline has water flowing at a velocity of 2.6 fps. What is the gpm flow rate through the pipeline? Assume the pipe is flowing full. (Round to the nearest tenth.)

ANS_____

5. A pump discharges into a 2-ft diameter barrel. If the water level in the barrel rises 26 inches in 30 seconds, what is the gpm flow into the barrel?

ANS_____

6. A 10-inch diameter pipeline has water flowing at a velocity of 3.2 fps. What is the gpm flow rate through the pipeline if the water is flowing at a depth of 5 inches?

ANS_____

PRACTICE PROBLEMS 2.2: Velocity Calculations

1. A channel has a rectangular cross section. The channel is 5 ft wide with water flowing to a depth of 2.3 ft. If the flow rate through the channel is 13,400 gpm, what is the velocity of the water in the channel (ft/sec)? (Round to the nearest tenth.)

                                                              ANS_____

2. An 8-inch diameter pipe flowing full delivers 537 gpm. What is the velocity of flow in the pipeline (ft/sec)? (Round to the nearest tenth.)

                                                              ANS_____

3. A fluorescent dye is used to estimate the velocity of flow in a sewer. The dye is injected into the water at one manhole and the travel time to the next manhole 500 ft away is noted. The dye first appears at the downstream manhole in 195 seconds. The dye continues to be visible until the total elapsed time is 221 seconds. What is the ft/sec velocity of flow through the pipeline? (Round to the nearest tenth.)

                                                              ANS_____

4. The velocity in a 10-inch diameter pipeline is 2.6 ft/sec. If the 10-inch pipeline flows into an 8-inch diameter pipeline, what is the velocity in the 8-inch pipeline in ft/sec.?

ANS_____

5. A float travels 400 ft in a channel in 1 min 28 sec. What is the estimated velocity in the channel (ft/sec)? (Round to the nearest tenth.)

ANS_____

6. The velocity in a 8-inch diameter pipe is 3.6 ft/sec. If the flow then travels through a 10-inch diameter section of pipeline, what is the ft/sec velocity in the 10-inch pipeline? (Round to the nearest tenth.)

ANS_____

PRACTICE PROBLEMS 2.3: Average Flow Rates

1. The following flows were recorded for the week: Monday—4.6 MGD; Tuesday—5.2 MGD; Wednesday—5.3 MGD; Thursday—4.9 MGD; Friday—5.4 MGD; Saturday—5.1 MGD; Sunday—4.8 MGD. What was the average daily flow rate for the week?

ANS_____

2. The totalizer reading for the month of November was 117.3 MG. What was the average daily flow (ADF) for the month of November? (Round to the nearest tenth.)

ANS_____

3. The following flows were recorded for the months of April, May, and June: April—125.6 MG; May—142.4 MG; June—160.2 MG. What was the average daily flow for this three-month period? (Round to the nearest tenth.)

ANS_____

4. The total flow for one day at a plant was 3,140,000 gallons. What was the average gpm flow for that day?

ANS_____

PRACTICE PROBLEMS 2.4: Flow Conversions

1. Express a flow of 5 cfs in terms of gpm.

ANS_____

2. What is 38 gps expressed as gpd?

ANS_____

3. Convert a flow of 4,270,000 gpd to cfm.

ANS_____

4. What is 5.6 MGD expressed as cfs? (Round to the nearest tenth.)

ANS_____

5. Express 423,690 cfd as gpm.

ANS_____

6. Convert 2730 gpm to gpd.

ANS_____

Chapter 2—Achievement Test

1. A channel has a rectangular cross section. The channel is 6 ft wide with water flowing to a depth of 2.6 ft. If the flow rate through the channel is 15,500 gpm, what is the velocity of the water in the channel (ft/sec)? (Round to the nearest tenth.)

ANS_____

2. The following flows were recorded for the week: Monday—4.1 MGD; Tuesday—3.4 MGD; Wednesday—3.9 MGD; Thursday—4.6 MGD; Friday—3.2 MGD; Saturday—4.9 MGD; Sunday—3.7 MGD. What was the average daily flow rate for the week?

ANS_____

3. A channel 50 inches wide has water flowing to a depth of 3.2 ft. If the velocity of the water is 3.9 fps, what is the cfm flow in the channel?

ANS_____

4. The following flows were recorded for the months of June, July, and August: June—105.2 MG; July—129.6 MG; August—142.8 MG. What was the average daily flow for this three-month period? (Round to the nearest tenth.)

ANS_____

5. A tank is 10 ft by 10 ft. With the discharge valve closed, the influent to the tank causes the water level to rise 8 inches in one minute. What is the gpm flow to the tank?

ANS_____

6. An 8-inch diameter pipe flowing full delivers 490 gpm. What is the ft/sec velocity of flow in the pipeline?

ANS_____

7. Express a flow of 8 cfs in terms of gpm.

ANS_____

8. The totalizer reading for the month of October was 127.6 MG. What was the average daily flow (ADF) for the month of October? (Round to the nearest tenth.)

ANS_____

Chapter 2—Achievement Test Cont'd

9. What is 4.8 MGD expressed as cfs? (Round to the nearest tenth.)

ANS_____

10. A pump discharges into a 2-ft diameter barrel. If the water level in the barrel rises 18 inches in 30 seconds, what is the gpm flow into the barrel?

ANS_____

11. Convert a flow of 1,780,000 gpd to cfm.

ANS_____

12. A 6-inch diameter pipeline has water flowing at a velocity of 2.7 fps. What is the gpm flow rate through the pipeline? (Assume the pipe is flowing full.)

ANS_____

13. A fluorescent dye is used to estimate the velocity of flow in a sewer. The dye is injected into the water at one manhole and the travel time to the next manhole 300 ft away is noted. The dye first appears at the downstream manhole in 77 seconds. The dye continues to be visible until there is a total elapsed time of 95 seconds. What is the ft/sec velocity of flow through the pipeline?

ANS_____

14. The velocity in a 10-inch pipeline is 2.4 ft/sec. If the 10-inch pipeline flows into an 8-inch diameter pipeline, what is the ft/sec velocity in the 8-inch pipeline? (Round to the nearest tenth.)

ANS_____

15. Convert 2150 gpm to gpd.

ANS_____

16. The total flow for one day at a plant was 4,620,000 gallons. What was the average gpm flow for that day?

ANS_____

3 *Milligrams Per Liter to Pounds Per Day Calculations*

PRACTICE PROBLEMS 3.1: Chemical Dosage Calculations

1. Determine the chlorinator setting (lbs/day) needed to treat a flow of 5.1 MGD with a chlorine dose of 2.3 mg/L.

ANS_____

2. To dechlorinate a wastewater, sulfur dioxide is to be applied at a level 3 mg/L more than the chlorine residual. What should the sulfonator feed rate be (lbs/day) for a flow of 3.8 MGD with a chlorine residual of 2.9 mg/L?

ANS_____

3. A total chlorine dosage of 8 mg/L is required to treat a particular water. If the flow is 1.6 MGD and the hypochlorite has 65% available chlorine, how many lbs/day of hypochlorite will be required?

ANS_____

4. What should the chlorinator setting be (lbs/day) to treat a flow of 4.6 MGD if the chlorine demand is 8.5 mg/L and a chlorine residual of 2 mg/L is desired?

ANS_____

5. The chlorine dosage at a plant is 4.1 mg/L. If the flow rate is 6,140,000 gpd, what is the chlorine feed rate in lbs/day?

ANS_____

6. A storage tank is to be disinfected with 50 mg/L of chlorine. If the tank holds 85,000 gallons, how many pounds of chlorine (gas) will be needed?

ANS_____

7. To neutralize a sour digester, one pound of lime is to be added for every pound of volatile acids in the digester liquor. If the digester contains 224,000 gal of sludge with a volatile acid (VA) level of 2,120 mg/L, how many pounds of lime should be added?

ANS_____

8. A flow of 0.72 MGD requires a chlorine dosage of 8 mg/L. If the hypochlorite has 65% available chlorine, how many lbs/day of hypochlorite will be required?

ANS_____

PRACTICE PROBLEMS 3.2: Loading Calculations—BOD, COD, and SS

1. The suspended solids concentration of the wastewater entering the primary system is 425 mg/*L*. If the plant flow is 1,620,000 gpd, how many lbs/day suspended solids enter the primary system?

ANS_____

2. Calculate the BOD loading (lbs/day) on a stream if the secondary effluent flow is 2.98 MGD and the BOD of the secondary effluent is 26 mg/*L*.

ANS_____

3. The daily flow to a trickling filter is 5,340,000 gpd. If the BOD content of the trickling filter influent is 280 mg/*L*, how many lbs/day BOD enter the trickling filter?

ANS_____

4. The flow to an aeration tank is 2460 gpm. If the COD concentration of the water is 135 mg/*L*, how many pounds of COD are applied to the aeration tank daily? (Round the MGD flow to the nearest hundredth.)

ANS_____

5. The daily flow to a trickling filter is 2290 gpm with a BOD concentration of 295 mg/*L*.. How many lbs of BOD are applied to the trickling filter daily? (Round the MGD flow to the nearest hundredth.)

ANS_____

PRACTICE PROBLEMS 3.3: BOD and SS Removal, lbs/day

1. If 148 mg/*L* suspended solids are removed by a primary clarifier, how many lbs/day suspended solids are removed when the flow is 5.2 MGD?

ANS_____

2. The flow to a primary clarifier is 1.89 MGD. If the influent to the clarifier has a suspended solids concentration of 315 mg/*L* and the primary effluent has 126 mg/*L* SS, how many lbs/day suspended solids are removed by the clarifier?

ANS_____

3. The flow to a trickling filter is 4,790,000 gpd. If the primary effluent has a BOD concentration of 160 mg/*L* and the trickling filter effluent has a BOD concentration of 24 mg/*L*, how many pounds of BOD are removed daily?

ANS_____

4. A primary clarifier receives a flow of 2.37 MGD with a suspended solids concentration of 387 mg/L. If the clarifier effluent has a suspended solids concentration of 166 mg/L, how many pounds of suspended solids are removed daily?

ANS_____

5. The flow to the trickling filter is 4,140,000 gpd with a BOD concentration of 215 mg/L. If the trickling filter effluent has a BOD concentration of 97 mg/L, how many lbs/day BOD are removed by the trickling filter?

ANS_____

PRACTICE PROBLEMS 3.4: Pounds of Solids Under Aeration

1. An aeration tank has a volume of 350,000 gallons. If the mixed liquor suspended solids concentration is 2140 mg/L, how many pounds of suspended solids are in the aerator?

ANS_____

2. The aeration tank of a conventional activated sludge plant has a mixed liquor volatile suspended solids concentration of 1960 mg/L. If the aeration tank is 110 ft long, 35 ft wide, and has wastewater to a depth of 13 ft, how many pounds of MLVSS are under aeration? (Round tank volume to the nearest ten thousand.)

ANS_____

3. The volume of an oxidation ditch is 24,500 cubic feet. If the MLVSS concentration is 2960 mg/L, how many pounds of volatile solids are under aeration? (Round ditch volume to the nearest ten thousand.)

ANS_____

4. An aeration tank is 120 ft long and 45 ft wide. The operating depth is 15 ft. If the mixed liquor suspended solids concentration is 2440 mg/*L*, how many pounds of mixed liquor suspended solids are under aeration? (Round tank volume to the nearest ten thousand.)

ANS_____

5. An aeration tank is 110 ft long and 40 ft wide. The depth of wastewater in the tank is 15 ft. If the tank contains an MLSS concentration of 2890 mg/*L*, how many lbs of MLSS are under aeration? (Round tank volume to the nearest ten thousand.)

ANS_____

PRACTICE PROBLEMS 3.5: WAS Pumping Rate Calculations

1. The WAS suspended solids concentration is 6210 mg/*L*. If 5300 lbs/day solids are to be wasted, what must the WAS pumping rate be, in MGD? (Round MGD flow to the nearest hundredth.)

ANS_____

2. The WAS suspended solids concentration is 5970 mg/*L*. If 4600 lbs/day solids are to be wasted, (a) What must the WAS pumping rate be, in MGD? (Round MGD flow to the nearest hundredth.) (b) What is this rate expressed in gpm?

ANS_____

3. It has been determined that 6090 lbs/day of solids must be removed from the secondary system. If the RAS SS concentration is 6540 mg/*L*, what must be the WAS pumping rate, in gpm? (Round MGD flow to the nearest hundredth.)

ANS_____

4. The RAS suspended solids concentration is 6280 mg/*L*. If a total of 7400 lbs/day solids are to be wasted, what should the WAS pumping rate be, in gpm? (Round MGD flow to the nearest hundredth.)

ANS_____

5. A total of 5700 lbs/day of solids must be removed from the secondary system. If the RAS SS concentration is 7140 mg/*L*, what must be the WAS pumping rate, in gpm? (Round MGD flow to the nearest hundredth.)

ANS_____

Chapter 3—Achievement Test

1. Determine the chlorinator setting (lbs/day) required to treat a flow of 3,820,000 gpd with a chlorine dose of 2.4 mg/L.

ANS_____

2. Calculate the BOD loading (lbs/day) on a stream if the secondary effluent flow is 2.05 MGD and the BOD of the secondary effluent is 15 mg/L.

ANS_____

3. The flow to a primary clarifier is 4.6 MGD. If the influent to the clarifier has a suspended solids concentration of 307 mg/L and the primary effluent suspended solids concentration is 122 mg/L, how many lbs/day suspended solids are removed by the clarifier?

ANS_____

4. What should the chlorinator setting be (lbs/day) to treat a flow of 5.6 MGD if the chlorine demand is 7.9 mg/L and a chlorine residual of 2 mg/L is desired?

ANS_____

5. The suspended solids concentration of the wastewater entering the primary system is 315 mg/*L*. If the plant flow is 3.7 MGD, how many lbs/day suspended solids enter the primary system?

ANS_____

6. A total chlorine dosage of 12 mg/*L* is required to treat a particular water. If the flow is 2.8 MGD and the hypochlorite has 65% available chlorine, how many lbs/day of hypochlorite will be required?

ANS_____

7. A primary clarifier receives a flow of 3.22 MGD with a suspended solids concentration of 340 mg/*L*. If the clarifier effluent has a suspended solids concentration of 150 mg/*L*, how many pounds of suspended solids are removed daily?

ANS_____

8. A storage tank is to be disinfected with 50 mg/*L* of chlorine. If the tank holds 80,000 gallons, how many pounds of chlorine gas will be needed?

ANS_____

Chapter 3—Achievement Test Cont'd

9. An aeration tank is 100 ft long and 40 ft wide. The operating depth is 13 ft. If the mixed liquor suspended solids concentration is 2610 mg/L, how many pounds of mixed liquor suspended solids are under aeration? (Round the tank volume to the nearest ten thousand.)

ANS_____

10. The WAS suspended solids concentration is 5980 mg/L. If 5540 lbs/day solids are to be wasted, what must the WAS pumping rate be, in MGD? (Round to the nearest hundredth MGD.)

ANS_____

11. The flow to an aeration tank is 2200 gpm. If the COD concentration of the water is 125 mg/L, how many pounds COD enter the aeration tank daily?

ANS_____

12. The daily flow to a trickling filter is 2190 gpm. If the BOD concentration of the trickling filter influent is 235 mg/L, how many lbs/day BOD are applied to the trickling filter?

ANS_____

13. The 1.9 MGD influent to the secondary system has a BOD concentration of 230 mg/*L*. The secondary effluent contains 25 mg/*L* BOD. How many pounds of BOD are removed each day by the secondary system?

ANS_____

14. The chlorine feed rate at a plant is 340 lbs/day. If the flow is 5,100,000 gpd, what is this dosage expressed in mg/*L*?

ANS_____

15. It has been determined that 6240 lbs/day solids must be removed from the secondary system. If the RAS SS concentration is 5800 mg/*L*, what must be the WAS pumping rate, in gpm? (Round the MGD flow to the nearest hundredth.)

ANS_____

4 Loading Rate Calculations

PRACTICE PROBLEMS 4.1: Hydraulic Loading Rate Calculations

1. A trickling filter 90 ft in diameter treats a primary effluent flow of 2.3 MGD. If the recirculated flow to the clarifier is 0.7 MGD, what is the hydraulic loading on the trickling filter in gpd/sq ft?

ANS_____

2. The flow to an 80-ft diameter trickling filter is 2,670,000 gpd. The recirculated flow is 1,650,000 gpd. At this flow rate, what is the hydraulic loading rate in gpd/sq ft?

ANS_____

3. A rotating biological contactor treats a flow of 3.6 MGD. The manufacturer data indicates a media surface area of 850,000 sq ft. What is the hydraulic loading rate on the RBC in gpd/sq ft? (Round answer to the nearest tenth.)

ANS_____

4. A pond receives a flow of 1,980,000 gpd. If the surface area of the pond is 15 acres, what is the hydraulic loading in in./day? (Round answer to the nearest tenth.)

ANS_____

5. What is the hydraulic loading rate in gpd/sq ft to an 85-ft diameter trickling filter if the primary effluent flow to the trickling filter is 3,780,000 gpd, and the recirculated flow is 1,300,000 gpd?

ANS_____

6. A 20-acre pond receives a flow of 4.3 acre-feet/day. What is the hydraulic loading on the pond in in./day?

ANS_____

PRACTICE PROBLEMS 4.2: Surface Overflow Rate Calculations

1. A sedimentation tank 80 ft by 30 ft receives a flow of 2.14 MGD. What is the surface overflow rate in gpd/sq ft?

ANS_____

2. A circular clarifier has a diameter of 70 ft. If the primary effluent flow is 2.62 MGD, what is the surface overflow rate in gpd/sq ft?

ANS_____

3. A sedimentation tank is 100 ft long and 40 ft wide. If the flow to the tank is 3.28 MGD what is the surface overflow rate in gpd/sq ft?

ANS_____

4. The primary effluent flow to a clarifier is 1.48 MGD. If the sedimentation tank is 20 ft wide and 80 ft long, what is the surface overflow rate of the clarifier in gpd/sq ft?

ANS_____

5. The flow to a circular clarifier is 2.36 MGD. If the diameter of the clarifier is 60 ft, what is the surface overflow rate in gpd/sq ft?

ANS_____

PRACTICE PROBLEMS 4.3: Filtration Rate Calculations

1. A filter 30 ft by 25 ft receives a flow of 2150 gpm. What is the filtration rate in gpm/sq ft? (Round to the nearest tenth.)

ANS_____

2. A filter 40 ft by 20 ft receives a flow rate of 3080 gpm. What is the filtration rate in gpm/sq ft?

ANS_____

3. A filter 25 ft by 50 ft receives a flow of 2000 gpm. What is the filtration rate in gpm/sq ft?

ANS_____

4. A filter 45 ft by 25 ft treats a flow of 2.2 MGD. What is the filtration rate in gpm/sq ft? (Round to the nearest tenth.)

ANS_____

5. A filter has a surface area of 875 sq ft. If the flow treated is 2975 gpm, what is the filtration rate in gpm/sq ft?

ANS_____

PRACTICE PROBLEMS 4.4: Backwash Rate Calculations

1. A filter 15 ft by 15 ft has a backwash flow rate of 4950 gpm. What is the filter backwash rate in gpm/sq ft?

ANS_____

2. A filter 25 ft by 15 ft has a backwash flow rate of 5100 gpm.. What is the filter backwash rate in gpm/sq ft?

ANS_____

3. A filter is 20 ft by 15 ft. If the backwash flow rate is 3300 gpm, what is the filter backwash rate in gpm/sq ft?

ANS_____

4. A filter 20 ft by 30 ft backwashes at a rate of 3200 gpm. What is this backwash rate expressed as gpm/sq ft?

ANS_____

5. The backwash flow rate for a filter is 3800 gpm. If the filter is 20 ft by 20 ft, what is the backwash rate expressed as gpm/sq ft?

ANS_____

PRACTICE PROBLEMS 4.5: Unit Filter Run Volume Calculations

1. The total water filtered during a filter run is 3,890,000 gallons. If the filter is 15 ft by 40 ft, what is the unit filter run volume (UFRV) in gal/sq ft?

ANS_____

2. The total water filtered during a filter run (between backwashes) is 1,680,000 gallons. If the filter is 15 ft by 15 ft, what is the UFRV in gal/sq ft?

ANS_____

3. A filter 20 ft by 25 ft filters a total of 3.96 MG during the filter run. What is the unit filter run volume in gal/sq ft?

ANS_____

4. The total water filtered between backwashes is 1,339,200 gal. If the length of the filter is 15 ft and the width is 12 ft, what is the unit filter run volume in gal/sq ft?

ANS_____

5. A filter is 30 ft by 25 ft. If the total water filtered between backwashes is 5,625,000 gallons, what is the UFRV in gal/sq ft?

ANS_____

PRACTICE PROBLEMS 4.6: Weir Overflow Rate Calculations

1. A rectangular clarifier has a total of 157 ft of weir. What is the weir overflow rate in gpd/ft when the flow is 1,397,000 gpd?

ANS_____

2. A circular clarifier receives a flow of 2.32 MGD. If the diameter of the weir is 60 ft, what is the weir overflow rate in gpd/ft ?

ANS_____

3. A rectangular clarifier has a total of 235 ft of weir. What is the weir overflow rate in gpd/ft when the flow is 2.6 MGD?

ANS_____

4. The flow rate to a clarifier is 1200 gpm. If the diameter of the weir is 70 ft, what is the weir overflow rate in gpd/ft ?

ANS_____

5. A rectangular sedimentation basin has a total weir length of 188 ft. If the flow to the basin is 4.06 MGD, what is the weir loading rate in gpm/ft?

ANS_____

PRACTICE PROBLEMS 4.7: Organic Loading Rate Calculations

1. A trickling filter 80 ft in diameter with a media depth of 5 feet receives a flow of 2,240,000 gpd. If the BOD concentration of the primary effluent is 215 mg/*L*, what is the organic loading on the trickling filter in lbs BOD/day/1000 cu ft? (Round media volume to nearest hundred.)

ANS_____

2. The flow to a 3.7-acre wastewater pond is 122,000 gpd. The influent BOD concentration is 175 mg/*L*. What is the organic loading to the pond in lbs BOD/day/ac?

ANS_____

3. An 85-ft diameter trickling filter with a media depth of 6 ft receives a primary effluent flow of 2,960,000 gpd with a BOD of 125 mg/*L*. What is the organic loading on the trickling filter in lbs BOD/day/1000 cu ft? (Round media volume to the nearest hundred.)

ANS_____

4. A rotating biological contactor (RBC) receives a flow of 2.15 MGD. If the soluble BOD of the influent wastewater to the RBC is 130 mg/*L* and the surface area of the media is 800,000 sq ft, what is the organic loading rate in lbs BOD/day/1000 sq ft?

ANS_____

5. A 90-ft diameter trickling filter with a media depth of 4 ft receives a primary effluent flow of 3.4 MGD. If the BOD concentration of the wastewater flow to the trickling filter is 140 mg/*L*, what is the organic loading rate in lbs BOD/day/1000 cu ft?(Round media volume to nearest hundred.)

ANS_____

PRACTICE PROBLEMS 4.8: Food/Microorganism Ratio Calculations

1. An activated sludge aeration tank receives a primary effluent flow of 3,350,000 gpd with a BOD of 207 mg/*L*. The mixed liquor volatile suspended solids is 1950 mg/*L* and the aeration tank volume is 400,000 gallons. What is the current F/M ratio?(Round to the nearest tenth.)

ANS_____

2. The volume of an aeration tank is 270,000 gallons. The mixed liquor suspended solids is 1690 mg/*L*. If the aeration tank receives a primary effluent flow of 3,150,000 gpd with a BOD of 192 mg/*L*, what is the F/M ratio? (Round to the nearest tenth.)

ANS_____

3. The desired F/M ratio at a particular activated sludge plant is 0.7 lbs COD/1 lb mixed liquor volatile suspended solids. If the 2.26 MGD primary effluent flow has a COD of 147 mg/*L* how many lbs of MLVSS should be maintained?

ANS_____

4. An activated sludge plant receives a flow of 1,920,000 gpd with a COD concentration of 155 mg/*L*. The aeration tank volume is 245,000 gallons and the MLVSS is 1880 mg/*L*. What is the current F/M ratio?(Round to the nearest tenth.)

ANS_____

5. The flow to an aeration tank is 2,940,000 gpd, with a BOD content of 185 mg/*L*. If the aeration tank is 100 ft long, 40 ft wide, has wastewater to a depth of 15 ft, and the desired F/M ratio is 0.5, what is the desired MLVSS concentration (mg/*L*) in the aeration tank? (Round tank volume to the nearest ten thousand.)

ANS'_____

PRACTICE PROBLEMS 4.9: Solids Loading Rate Calculations

1. A secondary clarifier is 80 ft in diameter and receives a combined primary effluent (P.E.) and return activated sludge (RAS) flow of 3.75 MGD. If the MLSS concentration in the aerator is 2780 mg/*L*, what is the solids loading rate on the secondary clarifier in lbs/day/sq ft? (Round to the nearest tenth.)

ANS_____

2. A secondary clarifier, 75 ft in diameter, receives a primary effluent flow of 2.95 MGD and a return sludge flow of 1.12 MGD. If the MLSS concentration is 2950 mg/*L*, what is the solids loading rate on the clarifier in lbs/day/sq ft? (Round to the nearest tenth.)

ANS_____

3. The desired solids loading rate for a 60-ft diameter clarifier is 28 lbs/day/sq ft. If the total flow to the clarifier is 3,590,000 gpd (P.E. + RAS flows), what is the desired MLSS concentration?

ANS_____

4. A secondary clarifier 50 ft in diameter receives a primary effluent flow of 2,450,000 gpd and a return sludge flow of 750,000 gpd. If the MLSS concentration is 2180 mg/*L*, what is the solids loading rate on the clarifier in lbs/day/sq ft? (Round to the nearest tenth.)

ANS_____

5. The desired solids loading rate for a 55-ft diameter clarifier is 20 lbs/day/sq ft. If the total flow to the clarifier is 2,960,000 gpd (P.E. + RAS flows), what is the desired MLSS concentration?

ANS_____

PRACTICE PROBLEMS 4.10: Digester Loading Rate Calculations

1. A digester receives a total of 11,650 lbs/day volatile solids. If the digester volume is 32,600 cu ft, what is the digester loading in lbs VS added/day/cu ft? (Round to the nearest hundredth.)

ANS_____

2. A digester 50 ft in diameter with a water depth of 20 ft receives 122,000 lbs/day raw sludge. If the sludge contains 6.5% solids with 70% volatile matter, what is the digester loading in lbs VS added/day/cu ft? (Round to the nearest hundredth.)

ANS_____

3. A digester 45 ft in diameter with a liquid level of 19 ft receives 139,000 lbs/day sludge with 6% total solids and 69% volatile solids. What is the digester loading in lbs VS added/day/cu ft? (Round to the nearest hundredth.)

ANS_____

4. A digester 35 ft in diameter with a liquid level of 15 ft receives 19,200 gpd sludge with 5.3% solids and 68% volatile solids. What is the digester loading in lbs VS/day/cu ft? Assume the sludge weighs 8.34 lbs/gal. (Round to the nearest hundredth.)

ANS_____

5. A digester 40 ft in diameter with a liquid level of 19 ft receives 21,000 gpd sludge with 5.2% total solids and 71% volatile solids. What is the digester loading in lbs VS/day/cu ft? Assume the sludge weighs 8.8 lbs/gal. (Round to the nearest hundredth.)

ANS_____

PRACTICE PROBLEMS 4.11: Digester Volatile Solids Loading Ratio Calculations

1. A total of 1930 lbs/day volatile solids are pumped to a digester. The digester sludge contains a total of 31,200 lbs of volatile solids. What is the volatile solids loading on the digester in lbs VS added/day/lb VS in digester? (Round to the nearest hundredth.)

ANS_____

2. A digester contains a total of 172,700 lbs of sludge that has a total solids content of 5.8% and volatile solids of 65%. If 550 lbs/day volatile solids are added to the digester, what is the volatile solids loading on the digester in lbs VS added/day/lb VS in digester? (Round to the nearest hundredth.)

ANS_____

3. A total of 61,200 lbs/day sludge is pumped to a 110,000-gallon digester. The sludge being pumped to the digester has a total solids content of 5.4% and a volatile solids content of 71%. The sludge in the digester has a solids content of 6.5% with a 57% volatile solids content. What is the volatile solids loading on the digester in lbs VS added/day/lb VS in digester? Assume the sludge in the digester weighs 8.34 lbs/gal. (Round to the nearest hundredth.)

ANS_____

4. A total of 108,000 gal of digested sludge is in a digester. The digested sludge contains 5.8% total solids and 56% volatile solids. If the desired VS loading ratio is 0.08 lbs VS added/day/lb VS under digestion, what is the desired lbs VS/day to enter the digester? Assume the sludge in the digester weighs 8.34 lbs/gal. (Round to the nearest hundredth.)

ANS_____

5. A total of 7,700 gpd sludge is pumped to the digester. The sludge has 4.4% solids with a volatile solids content of 72%. If the desired VS loading ratio is 0.06 lbs VS added/day/lb VS under digestion, how many lbs VS should be in the digester for this volatile solids load? Assume the sludge pumped to the digester weighs 8.34 lbs/gal. (Round to the nearest hundredth.)

ANS_____

PRACTICE PROBLEMS 4.12: Population Loading and Population Equivalent

1. A 4.7-acre wastewater pond serves a population of 1530. What is the population loading on the pond in persons per acre?

ANS_____

2. A wastewater pond serves a population of 3825. If the pond is 9 acres, what is the population loading on the pond?

ANS_____

3. A 372,000-gpd wastewater flow has a BOD concentration of 1710 mg/*L*. Using an average of 0.2 lbs/day BOD/person, what is the population equivalent of this wastewater flow?

ANS_____

4. A wastewater pond is designed to serve a population of 5000. If the desired population loading is 350 people per acre, how many acres of pond will be required? (Round the answer to the nearest tenth.)

ANS_____

5. A 98,000-gpd wastewater flow has a BOD content of 2190 mg/*L*. Using an average of 0.2 lbs/day BOD/person, what is the population equivalent of this flow?

ANS_____

Chapter 4—Achievement Test

1. A circular clarifier has a diameter of 75 ft. If the primary effluent flow is 2.14 MGD, what is the surface overflow rate in gpd/sq ft?

ANS_____

2. A filter has a square foot area of 180 sq ft. If the flow rate to the filter is 2940 gpm, what is this filter backwash rate expressed as gpm/sq ft? (Round to the nearest tenth.)

ANS_____

3. The flow rate to a circular clarifier is 1,990,000 gpd. If the diameter of the weir is 75 ft, what is the weir overflow rate in gpd/ft?

ANS_____

4. A trickling filter, 80 ft in diameter, treats a primary effluent flow of 2.6 MGD. If the recirculated flow to the clarifier is 0.5 MGD, what is the hydraulic loading on the trickling filter in gpd/sq ft?

ANS_____

5. The desired F/M ratio at an activated sludge plant is 0.6 lbs BOD/day/lb mixed liquor volatile suspended solids. If the 1.8-MGD primary effluent flow has a BOD of 156 mg/*L*, how many lbs of MLVSS should be maintained in the aeration tank?

ANS_____

6. A digester contains a total of 175,000 lbs sludge that has a total solids content of 6.2% and volatile solids of 68%. If 400 lbs/day volatile solids are added to the digester, what is the volatile solids loading on the digester in lbs/day VS added/lb VS in the digester? (Round to the nearest hundredth.)

ANS_____

7. A secondary clarifier is 75 ft in diameter and receives a combined primary effluent (P.E.) and return activated sludge (RAS) flow of 3.45 MGD. If the MLSS concentration in the aerator is 2640 mg/*L*, what is the solids loading rate on the secondary clarifier in lbs/day/sq ft?

ANS_____

8. A digester, 60 ft in diameter with a water depth of 22 ft, receives 110,000 lbs/day raw sludge. If the sludge contains 6.8% solids with 70% volatile solids, what is the digester loading in lbs VS added/day/cu ft volume? (Round to the nearest hundredth.)

ANS_____

Chapter 4—Achievement Test Cont'd

9. A 20-acre pond receives a flow of 3.95 acre-feet/day. What is the hydraulic loading on the pond in in./day? (Round to the nearest tenth.)

ANS_____

10. The flow to an aeration tank is 3,154,000 gpd with a BOD content of 172 mg/*L*. If the aeration tank is 90 ft long, 40 ft wide, has wastewater to a depth of 14 ft, and the desired F/M ratio is 0.5, what is the desired MLVSS concentration (mg/*L*) in the aeration tank? (Round the tank volume to the nearest thousand.)

ANS_____

11. What is the solids loading rate for a 70-ft diameter clarifier in lbs/day/sq ft if the total flow to the clarifier is 4,100,000 (P.E. + RAS flows) and the MLSS concentration is 3180 mg/*L*?

ANS_____

12. A sedimentation tank 90 ft by 30 ft receives a flow of 2.1 MGD. What is the surface overflow rate in gpd/sq ft?

ANS_____

13. The total water filtered during a filter run (between backwashes) is 1,740,000 gallons, If the filter is 20 ft by 15 ft, what is the unit filter run volume (UFRV) in gallons/sq ft?

ANS_____

14. The volume of an aeration tank is 290,000 gallons. The mixed liquor volatile suspended solids is 1890 mg/*L*. If the aeration tank receives a primary effluent flow of 2,680,000 gpd with a COD of 140 mg/*L*, what is the F/M ratio? (Round to the nearest tenth.)

ANS_____

15. A total of 23,800 gallons of digested sludge is in a digester. The digested sludge contains 5.7% solids and 58% volatile solids. To maintain a desired VS loading ratio of 0.07 lbs VS added/day/lb VS under digestion, what is the desired lbs VS /day loading to the digester? (Assume sludge in the digester weighs 8.34 lbs/gal.)

ANS_____

16. The flow to a filter is 4.32 MGD. If the filter is 40 ft by 25 ft, what is the filter loading rate in gpm/sq ft?

ANS_____

Chapter 20—Achievement Test Cont'd

17. An 80-ft diameter trickling filter with a media depth of 4 ft receives a primary effluent flow of 3.1 MGD with a BOD concentration of 110 mg/L. What is the organic loading on the filter in lbs BOD/day/1000 cu ft?

ANS_____

18. A circular clarifier receives a flow of 2.46 MGD. If the diameter of the weir is 70 ft, what is the weir overflow rate in gpd/ft?

ANS_____

19. A 5.2-acre wastewater pond serves a population of 1800. What is the population loading on the pond (people/acre)?

ANS_____

20. A rotating biological contactor (RBC) receives a flow of 2.41 MGD. If the soluble BOD of the influent wastewater to the RBC is 120 mg/L and the surface area of the media is 750,000 sq ft, what is the organic loading rate in lbs Sol. BOD/day/1000 sq ft?

ANS_____

21. A filter 40 ft by 20 ft treats a flow of 2.72 MGD. What is the filter loading rate in gpm/sq ft?

ANS_____

5 *Detention and Retention Times Calculations*

PRACTICE PROBLEMS 5.1: Detention Time Calculations

1. A flocculation basin is 7 ft deep, 15 ft wide, and 30 ft long. If the flow through the basin is 1.35 MGD, what is the detention time in minutes?

ANS_____

2. The flow to a sedimentation tank 75 ft long, 25 ft wide, and 10 ft deep is 1.6 MGD. What is the detention time in the tank in hours?

ANS_____

3. A basin, 4 ft by 5 ft, is to be filled to the 2.5-ft level. If the flow to the tank is 5 gpm, how long will it take to fill the tank (in hours)? (Round to the nearest tenth.)

ANS_____

4. The flow rate to a circular clarifier is 4.87 MGD. If the clarifier is 70 ft in diameter with water to a depth of 12 ft, what is the detention time, in hours? (Round to the nearest tenth.)

ANS_____

5. A waste treatment pond is operated at a depth of 5 feet. The average width of the pond is 400 ft and the average length is 550 ft. If the flow to the pond is 219,400 gpd, what is the detention time, in days?

ANS_____

PRACTICE PROBLEMS 5.2: Sludge Age Calculations

1. An aeration tank has a total of 11,900 lbs of mixed liquor suspended solids. If a total of 2627 lbs/day suspended solids enter the aerator in the primary effluent flow, what is the sludge age in the aeration tank? (Round to the nearest tenth.)

ANS_____

2. An aeration tank is 100 ft long, 25 ft wide with wastewater to a depth of 15 ft. The mixed liquor suspended solids concentration is 2670 mg/L. If the primary effluent flow is 958,000 gpd with a suspended solids concentration of 128 mg/L, what is the sludge age in the aeration tank? (Round to the nearest tenth.)

ANS_____

3. An aeration tank contains 220,000 gallons of wastewater. The MLSS is 2960 mg/L. If the primary effluent flow is 1.46 MGD with a suspended solids concentration of 84 mg/L, what is the sludge age?

ANS_____

4. The 1.25 MGD primary effluent flow to an aeration tank has a suspended solids concentration of 70 mg/L. The aeration tank volume is 195,000 gallons. If a sludge age of 6.5 days is desired, what is the desired MLSS concentration?

ANS_____

5. A sludge age of 5.5 days is desired. Assume 1560 lbs/day suspended solids enter the aeration tank in the primary effluent. To maintain the desired sludge age, how many lbs of MLSS must be maintained in the aeration tank?

ANS_____

PRACTICE PROBLEMS 5.3: Solids Retention Time Calculations

1. An aeration tank has a volume of 270,000 gallons. The final clarifier has a volume of 190,000 gallons. The MLSS concentration in the aeration tank is 3100 mg/L. If a total of 1540 lbs/day SS are wasted and 330 lbs/day SS are in the secondary effluent, what is the solids retention time for the activated sludge system? Use the SRT equation that uses combined aeration tank and final clarifier volumes to estimate system solids. (Round to the nearest tenth.)

ANS_____

2. Determine the solids retention time (SRT) given the data below. Use the SRT equation that uses combined aeration tank and final clarifier volumes to estimate system solids. (Round answer to the nearest tenth.)

Aer. Vol. 225,000 gal   MLSS 2810 mg/L
Fin. Clar. Vol. 100,000 gal   WAS SS 5340 mg/L
P.E. Flow 2.15 MGD   S.E. SS 15 mg/L
WAS Pumping Rate 18,500 gpd

ANS_____

3. Calculate the solids retention time given the data below. Use the SRT equation that uses combined aeration tank and final clarifier volumes to estimate system solids. (Round to the nearest tenth.)

> Aer. Tank Vol. 1.2 MG MLSS 2440 mg/*L*
> Fin. Clar. Vol. 0.3 MG WAS SS 6120 mg/*L*
> P.E. Flow 2.6 MGD S.E. SS 18 mg/*L*
> WAS Pumping Rate 75,000 gpd

ANS_____

4. The volume of an aeration tank is 600,000 gal and the final clarifier is 155,000 gal. The desired SRT for the plant is 8 days. The primary effluent flow is 2.4 MGD and the WAS pumping rate is 30,000 gpd. If the WAS SS concentration is 6320 mg/*L* and the secondary effluent SS concentration is 22 mg/*L*, what is the desired MLSS concentration in mg/*L*? (Use the SRT equation that uses combined aeration tank and final clarifier volumes to estimate system solids.)

ANS_____

Chapter 5—Achievement Test

1. The flow to a sedimentation tank 70 ft long, 25 ft wide and 12 ft deep is 1,580,000 gpd. What is the detention time in the tank in hours? (Round to the nearest tenth.)

ANS_____

2. An aeration tank has a total of 12,400 lbs of mixed liquor suspended solids. If a total of 2750 lbs/day suspended solids enter the aeration tank in the primary effluent flow, what is the sludge age in the aeration tank? (Round to the nearest tenth.)

ANS_____

3. An aeration tank has a volume of 290,000 gallons. The final clarifier has a volume of 180,000 gallons. The MLSS concentration in the aeration tank is 2950 mg/*L*. If a total of 1620 lbs/day suspended solids are wasted and 310 lbs/day suspended solids are in the secondary effluent, what is the solids retention time for the activated sludge system? Use the combined aeration tank and clarifier volume to calculate system solids. (Round to the nearest tenth.)

ANS_____

4. The flow through a flocculation basin is 1.62 MGD. If the basin is 35 ft long, 15 ft wide, and 8 ft deep, what is the detention time in minutes?

ANS_____

5. Determine the solids retention time (SRT) given the data below. Use the combined aeration tank and clarifier volume to calculate system solids. (Round to the nearest tenth.)

Aer. Vol.—200,000 gal
Fin. Clar. Vol.—125,000 gal
P.E. Flow—2,300,000 gpd
WAS Pumping Rate—19,200 gpd

MLSS—2740 mg/*L*
WAS SS—5910 mg/*L*
SE SS—16 mg/*L*

ANS_____

6. The mixed liquor suspended solids concentration in an aeration tank is 3140 mg/*L*. The aeration tank contains 320,000 gallons. If the primary effluent flow is 2,240,000 gpd with a suspended solids concentration of 105 mg/*L*, what is the sludge age? (Round to the nearest tenth.)

ANS_____

Chapter 5—Achievement Test—Cont'd

7. Calculate the solids retention time given the following data:
Use the SRT equation that includes the combined aeration tank and secondary clarifier volumes to estimate system solids. (Round to the nearest tenth.)

Aer. Tank Vol.—1.3 MG   MLSS—2370 mg/L
Fin. Clar. Vol.—0.4 MG   WAS SS—6210 mg/L
P.E. Flow—2.72 MGD   SE SS—20 mg/L
WAS Pumping Rate—69,400 gpd

ANS_____

8. An aeration tank is 90 ft long, 30 ft wide, with wastewater to a depth of 12 ft. The mixed liquor suspended solids concentration is 2580 mg/L. If the influent flow to the aeration tank is 810,000 gpd with a suspended solids concentration of 145 mg/L, what is the sludge age in the aeration tank? (Round tank volume to the nearest ten thousand and round the sludge age answer to the nearest tenth.)

ANS_____

9. A tank 5 ft in diameter is to be filled to the 3-ft level. If the flow to the tank is 10 gpm, how long will it take to fill the tank (in min)?

ANS_____

10. A sludge age of 5 days is desired. The suspended solids concentration of the 1.38 MGD influent flow to the aeration tank is 135 mg/L. To maintain the desired sludge age, how many pounds of MLSS must be maintained in the aeration tank?

ANS_____

11. The average width of a pond is 300 ft and the average length is 480 ft. The depth is 5 ft. If the flow to the pond is 195,000 gpd, what is the detention time in days? (Round to the nearest tenth.)

ANS_____

12. The volume of an aeration tank is 540,000 gal and the volume of the final clarifier is 170,000 gal. The desired SRT for the plant is 8 days. The primary effluent flow is 2,830,000 gpd and the WAS pumping rate is 36,000 gpd. If the WAS SS concentration is 6240 mg/L, and the secondary effluent SS concentration is 15 mg/L, what is the desired MLSS concentration in mg/L? (Use the SRT equation that includes the combined aeration tank and secondary clarifier volumes to estimate system solids.)

ANS_____

6 *Efficiency and Other Percent Calculations*

PRACTICE PROBLEMS 6.1: Unit Process Efficiency Calculations

1. The suspended solids concentration entering a trickling filter is 120 mg/*L*. If the suspended solids concentration in the trickling filter effluent is 22 mg/*L*, what is suspended solids removal efficiency of the trickling filter?

ANS_____

2. The BOD concentration of the raw wastewater at an activated sludge plant is 245 mg/*L*. If the BOD concentration of the final effluent is 15 mg/*L*, what is the overall efficiency of the plant in BOD removal?

ANS_____

3. The influent flow to a waste treatment pond has a BOD content of 270 mg/*L*. If the pond effluent has a BOD content of 62 mg/*L*, what is the BOD removal efficiency of the pond?

ANS_____

4. The influent of a primary clarifier has a BOD content of 230 mg/*L*. If 95 mg/*L* BOD are removed, what is the BOD removal efficiency?

ANS_____

5. The suspended solids concentration of the primary clarifier influent is 305 mg/L. If the suspended solids concentration of the primary effluent is 137 mg/L, what is the suspended solids removal efficiency?

ANS_____

PRACTICE PROBLEMS 6.2: Percent Solids and Sludge Pumping Rate Calculations

1. A total of 3610 gallons of sludge are pumped to a digester. If the sludge has a 5.8% solids content, how many lbs/day solids are pumped to the digester? (Assume the sludge weighs 8.34 lbs/gal.)

ANS_____

2. The total weight of a sludge sample is 13.05 grams (sludge sample only, not the dish). If the weight of the solids after drying is 0.68 grams, what is the percent total solids of the sludge?

ANS_____

3. A total of 1430 lbs/day SS are removed from a primary clarifier and pumped to a sludge thickener. If the sludge has a solids content of 3.2%, how many lbs/day sludge is this?

ANS_____

4. It is anticipated that 265 lbs/day SS will be pumped from the primary clarifier of a new plant. If the primary clarifier sludge has a solids content of 4.4%, how many gpd sludge will be pumped from the clarifier? (Assume a sludge weight of 8.34 lbs/gal.)

ANS_____

5. A total of 286,000 lbs/day sludge is pumped from a primary clarifier to a sludge thickener. If the total solids content of the sludge is 3.4%, how many lbs/day total solids are sent to the thickener?

ANS_____

PRACTICE PROBLEMS 6.3: Mixing Different Percent Solids Sludges Calculations

1. A primary sludge flow of 3000 gpd with a solids content of 4.5% is mixed with a thickened secondary sludge flow of 4200 gpd that has a solids content of 3.8%. What is the percent solids content of the mixed sludge flow? Assume the density of both sludges is 8.34 lbs/gal. (Round to the nearest tenth percent.)

ANS_____

2. Primary and thickened secondary sludges are to be mixed and sent to the digester. The 8200-gpd primary sludge has a solids content of 5.2% and the 7000-gpd thickened secondary sludge has a solids content of 4.2%. What would be the percent solids content of the mixed sludge? (Assume the density of both sludges is 8.34 lbs/gal.)

ANS_____

3. A 4840-gpd primary sludge has a solids content of 4.9%. The 5200-gpd thickened secondary sludge has a solids content of 3.6%. If the sludges were blended, what would be the percent solids content of the mixed sludge? (Assume the density of both sludges is 8.34 lbs/gal.)

ANS_____

4. A primary sludge flow of 9010 gpd with a solids content of 4.2% is mixed with a thickened secondary sludge flow of 10,760 gpd with a 6.5% solids content. What is the percent solids of the combined sludge flow? (Assume the density of both sludges is 8.34 lbs/gal.)

ANS_____

PRACTICE PROBLEMS 6.4: Percent Volatile Solids Calculations

1. If 3340 lbs/day solids with a volatile solids content of 70% are sent to the digester, how many lbs/day volatile solids are sent to the digester?

ANS_____

2. A total of 4070 gpd of sludge is to be pumped to the digester. If the sludge has a 7% solids content with 72% volatile solids, how many lbs/day volatile solids are pumped to the digester? (Assume the sludge weighs 8.34 lbs/gal.)

ANS_____

3. How many lbs/day volatile solids are pumped to the digester if a total of 6400 gpd of sludge is to be pumped to the digester? The sludge has a 6.5% solids content of which 67% are volatile solids. (Assume the sludge weighs 8.34 lbs/gal.)

ANS_____

4. A 6.8% sludge has a volatile solids content of 66%. If 24,510 lbs/day of sludge are pumped to the digester, how many lbs/day of volatile solids are pumped to the digester?

ANS_____

5. A sludge has a solids content of 6.2% and a volatile solids content of 71%. If 2530 gpd of sludge are pumped to the digester, how many lbs/day volatile solids are pumped to the digester? (Assume the sludge weighs 8.34 lbs/gal.)

ANS_____

PRACTICE PROBLEMS 6.5: Seed Sludge Based on Percent Digester Volume Calculations

1. A digester has a capacity of 280,000 gallons. If the digester seed sludge is to be 18% of the digester capacity, how many gallons of seed sludge will be required?

ANS_____

2. A digester 70 ft in diameter has a side wall water depth of 21 ft. If the digester seed sludge is to be 20% of the digester capacity, how many gallons of seed sludge will be required?

ANS_____

3. A 50-ft diameter digester has a typical water depth of 21 ft. If the seed sludge to be used is 15% of the tank capacity, how many gallons of seed sludge will be required?

ANS_____

4. A 60-ft diameter digester has a typical side wall water depth of 23 ft. If 87,500 gallons of seed sludge are to be used in starting up the digester, what percent of the digester volume will be seed sludge?

ANS_____

PRACTICE PROBLEMS 6.6: Solution Strength Calculations

1. A total of 2 lbs of chemical is dissolved in 80 lbs of solution. What is the percent strength, by weight, of the solution?

ANS_____

2. If 6 ounces of dry polymer are added to 8 gallons of water, what is the percent strength (by weight) of the polymer solution? (Round pounds dry polymer and pounds water to the nearest tenth and round the answer to the nearest tenth.)

ANS_____

3. How many pounds of dry polymer must be added to 50 gallons of water to make a 1.5% (by weight) polymer solution? (Round the answer to the nearest tenth.)

ANS_____

4. If 500 grams of dry polymer are dissolved in 8 gallons of water, what percent strength is the solution? (1 gram = 0.0022 lbs)

ANS_____

5. How many grams of chemical must be dissolved in 5 gallons of water to make a 1.8% solution? (1 lb = 454 grams)

ANS_____

PRACTICE PROBLEMS 6.7: Mixing Different Percent Strength Solutions Calculations

1. If 15 lbs of a 9% strength solution are mixed with 100 lbs of a 1% strength solution, what is the percent strength of the solution mixture? (Round to the nearest tenth.)

ANS_____

2. If 10 lbs of a 12% strength solution are mixed with 330 lbs of a 0.3% strength solution, what is the percent strength of the solution mixture? (Round to the nearest tenth.)

ANS_____

3. If 20 lbs of an 12% strength solution is mixed with 445 lbs of a 0.5% strength solution, what is the percent strength of the solution mixture? (Round to the nearest tenth.)

ANS_____

4. If 10 gallons of a 11% strength solution are added to 60 gallons of a 0.1% strength solution, what is the percent strength of the solution mixture? Assume the 11% strength solution weighs 9.8 lbs/gal and the 0.1% strength solution weighs 8.34 lbs/gal. (Round to the nearest tenth.)

ANS_____

PRACTICE PROBLEMS 6.8: Pump and Motor Efficiency Calculations

1. The brake horsepower of a pump is 22 hp. If the water horsepower is 17 hp, what is the efficiency of the pump?

    ANS_____

2. If the motor horsepower is 50 hp and the brake horsepower is 43 hp, what is the efficiency of the motor?

    ANS_____

3. The motor horsepower is 25 hp. If the motor is 89% efficient, what is the brake horsepower?

    ANS_____

4. A total of 50 hp is supplied to a motor. If the wire-to-water efficiency of the pump and motor is 62%, what will the whp be?

ANS_____

5. The brake horsepower is 34.4 hp. If the motor is 86% efficient, what is the motor horsepower?

ANS_____

Chapter 6—Achievement Test

1. The BOD concentration of the raw wastewater at an activated sludge plant is 230 mg/*L*. If the BOD concentration of the final effluent is 21 mg/*L*, what is the overall efficiency of the plant in BOD removal?

    ANS_____

2. A total of 7200 gpd sludge is to be pumped to the digester. If the sludge has a 7% solids content with 68% volatile solids, how many lbs/day volatile solids are pumped to the digester? (Assume the sludge weighs 8.34 lbs/gal.)

    ANS_____

3. A primary sludge flow of 2600 gpd with a solids content of 4.2% is mixed with a secondary sludge flow of 3380 gpd with a solids content of 3.5%. What is the percent solids content of the mixed sludge flow? (Assume the weight of both sludges is 8.34 lbs/gal.)

    ANS_____

4. A total of 3 pounds of chemical has been dissolved in 92 pounds of solution. What is the percent strength, by weight, of the solution?

    ANS_____

5. A digester 80 ft in diameter has a side water depth of 23 ft. If the digester seed sludge is to be 20% of the digester capacity, how many gallons of seed sludge will be required?

ANS_____

6. If 12 lbs of a 10% strength solution are mixed with 400 lbs of a 0.15% strength solution, what is the percent strength of the solution mixture? (Round to the nearest tenth.)

ANS_____

7. The brake horsepower of a pump is 25.5 hp. If the water horsepower is 20.4 hp, what is the efficiency of the pump?

ANS_____

8. The influent flow to a waste treatment pond has a BOD content of 350 mg/*L*. If the pond effluent has a BOD content of 81 mg/*L*, what is the BOD removal efficiency of the pond?

ANS_____

Chapter 6—Achievement Test—Cont'd

9. A primary sludge flow of 8700 gpd with a solids content of 4.8% is mixed with a thickened secondary sludge flow of 11,200 gpd with a solids content of 6.2%. What is the percent solids of the combined sludge flow? Assume the weight of both sludges is 8.34 lbs/gal. (Round to the nearest tenth.)

ANS_____

10. How many lbs/day volatile solids are pumped to the digester if a total of 7400 gpd of sludge is pumped to the digester? The sludge has a 6.2% solids content, of which 72% are volatile solids.

ANS_____

11. A total of 35,400 lbs/day of sludge is pumped to the digester. If the sludge has a solids content of 5.7%, how many lbs/day solids are pumped to the digester?

ANS_____

12. How many pounds of dry polymer must be added to 75 gallons of water to make a 0.5% polymer solution?

ANS_____

13. A digester has a capacity of 190,000 gallons. If the digester seed sludge is to be 25% of the digester capacity, how many gallons of seed sludge will be required?

ANS_____

14. If 6800 lbs/day solids with a volatile solids content of 71% are pumped to the digester, how many lbs/day volatile solids are pumped to the digester?

ANS_____

15. If 8 pounds of a 9% strength solution are mixed with 90 pounds of a 1% strength solution, what is the percent strength of the solution mixture? (Round to the nearest tenth.)

ANS_____

16. If the motor horsepower is 50 hp and the brake horsepower is 38 hp, what is the efficiency of the motor?

ANS_____

17. A total of 75 hp is supplied to a motor. If the wire-to-water efficiency of the pump and motor is 67%, what will the water horsepower be?

ANS_____

7 *Pumping Calculations*

PRACTICE PROBLEMS 7.1: Density and Specific Gravity

1. A gallon of solution is weighed. After the weight of the container is subtracted, it is determined that the weight of the solution is 8.9 lbs. What is the density of the solution?

ANS_____

2. The density of a substance is given as 67.1 lbs/cu ft. What is this density expressed as lbs/gal?

ANS_____

3. The density of a liquid is given as 55 lbs/cu ft. What is the specific gravity of the liquid?

ANS_____

4. The specific gravity of a liquid is 1.3. What is the density of that liquid? (The density of water is 8.34 lbs/gal.)

ANS_____

5. You wish to determine the specific gravity of a solution. After weighing a gallon of solution and subtracting the weight of the container, the solution is found to weigh 8.9 lbs. What is the specific gravity of the solution?

ANS_____

PRACTICE PROBLEMS 7.2: Pressure and Force Calculations

1. The object shown below weighs 75 lbs. What is the lbs/sq in. pressure at the surface of contact? (Round to the nearest tenth.)

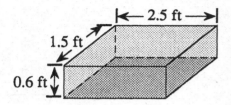

ANS_____

2. An object rests on the floor. The pressure at the surface of contact is 2.6 lbs/sq in. If the object is placed on another side that has only one-fourth the contact area, what is the new lbs/sq in. pressure? (Round to the nearest tenth.)

ANS_____

3. Compare the pressures, in lbs/sq ft, on the contact area for the two positions of the object shown below. The object weighs 210 lbs.

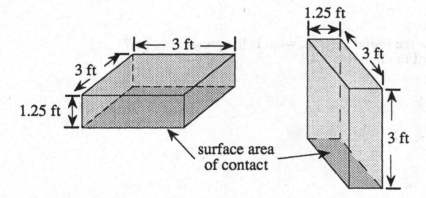

surface area
of contact

ANS_____

4. What is the pressure (in lbs/sq ft) at a point 6 ft below the surface of the water? (The density of water is 62.4 lbs/cu ft.)

ANS_____

5. What is the pressure (in psi) at a point 8 ft below the surface?

ANS_____

6. At a point 4.5 ft below the liquid surface, what is the pressure in psi? (The specific gravity of the liquid is 1.4.)

ANS_____

PRACTICE PROBLEMS 7.2: Pressure and Force Calculations—Cont'd

7. What is the total force against the bottom of a tank 20 ft long and 12 ft wide? The water depth is 10 ft. (Round psi to the nearest tenth.)

ANS_____

8. What is the total force exerted on the side of a tank if the tank is 25 ft wide and the water depth is 10 ft?

ANS_____

9. The side of a tank is 15 ft wide. The water depth is 9 ft. At what depth is the center of force against the tank wall?

ANS_____

10. The force applied to the small cylinder of a hydraulic jack is 50 lbs. The diameter of the small cylinder is 8 inches. If the diameter of the large cylinder is 2 ft, what is the total lifting force?

ANS_____

11. A gage reading is 32 psi. What is the absolute pressure at the gage? (Assume sea level atmospheric pressure.)

ANS_____

PRACTICE PROBLEMS 7.3: Head and Head Loss Calculations

1. The elevation of two water surfaces are 322 ft and 239 ft. What is the total static head, in ft?

ANS_____

2. The elevation of two water surfaces are 790 ft and 614 ft. If the friction and minor head losses equal 16 ft, what is the total dynamic head (in ft)?

ANS_____

3. The pump inlet and outlet pressure gage readings are given below. (Pump is off.) If the friction and minor head losses are equal to 12 ft, what is the total dynamic head (in ft)?

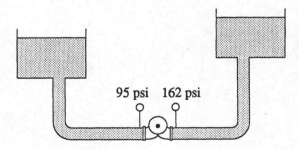

95 psi   162 psi

ANS_____

4. A 6-inch diameter pipe has a C-value of 100. When the flow rate is 900 gpm, what is the friction loss for a 1000-ft length of pipe? (Use the table given in Appendix A.)

ANS_____

5. Flow through a 10-inch diameter pipeline is 1450 gpm. The *C*-value is 100. What is the friction loss through a 2000-ft section of pipe? (Use the table given in Appendix A.)

ANS_____

6. Determine the "equivalent length of pipe" for a flow through a gate valve, 1/4 closed, for an 8-inch diameter pipeline. (Use the nomograph given in Appendix B.)

ANS_____

PRACTICE PROBLEMS 7.4: Horsepower Calculations

1. A pump must pump 1500 gpm against a total head of 40 ft. What horsepower is required for this work? (Round to the nearest tenth.)

ANS_____

2. If 20 hp is supplied to a motor (mhp), what is the bhp and whp if the motor is 85% efficient and the pump is 80% efficient? (Round to the nearest tenth.)

ANS_____

3. A total of 35 hp is required for a particular pumping application. If the pump efficiency is 85%, what is the brake horsepower required? (Round to the nearest tenth.)

ANS_____

4. A pump must pump against a total dynamic head of 60 ft at a flow rate of 600 gpm. The liquid to be pumped has a specific gravity of 1.2. What is the water horsepower requirement for this pumping application?

ANS_____

5. The motor horsepower requirement has been calculated to be 45 hp. How many kilowatts electric power does this represent? (1 hp = 746 watts)

ANS_____

6. The motor horsepower requirement has been calculated to be 75 hp. During the week, the pump is in operation a total of 144 hours. Using a power cost of $0.06125/kWh, what would be the power cost that week for the pumping?

ANS_____

PRACTICE PROBLEMS 7.5: Pump Capacity Calculations

1. A wet well is 12 ft long and 10 ft wide. The influent valve to the wet well is closed. If a pump lowers the water level 2.6 ft during a 5-minute pumping test, what is the gpm pumping rate?

ANS_____

2. A pump is discharged into a 55-gallon barrel. If it takes 29 seconds to fill the barrel, what is the pumping rate?

ANS_____

3. A pump test is conducted for 5 minutes while influent flow continues. During the test, the water level rises 2 inches. If the tank is 8 ft by 10 ft and the influent flow is 400 gpm, what is the pumping rate in gpm?

ANS_____

4. A piston pump discharges a total of 0.75 gal/stroke. If the pump operates at 30 strokes per minute, what is the gpm pumping rate? Assume the piston is 100% efficient and displaces 100% of its volume each stroke. (Round to the nearest tenth.)

ANS_____

5. A sludge pump has a bore of 8 inches and a stroke setting of 4 inches. The pump operates at 40 strokes per minute. If the pump operates a total of 150 minutes during a 24-hour period, what is the gpd pumping rate? Assume the piston is 100% efficient.

ANS_____

Chapter 7—Achievement Test

1. The density of a substance is given as 66 lbs/cu ft. What is this density expressed as lbs/gal?

ANS_____

2. The dimensions of a wet well are 12 ft by 10 ft. The influent valve to the wet well is closed. If a pump lowers the water level 1.5 ft during a 5-minute pumping test, what is the gpm capacity of the pump?

ANS_____

3. The object shown below weighs 90 lbs. What is the lbs/sq in. pressure at the surface of contact?

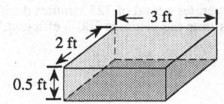

ANS_____

4. A sludge pump has a bore of 6 inches and a stroke of 2.5 inches. If the pump operates at 55 strokes (or revolutions) per minute, how many gpm are pumped? (Assume the piston is 100% efficient and displaces 100% of its volume each stroke.)

ANS_____

5. A pump test is conducted for 5 minutes. If the water level in the 12 ft long and 10 ft wide wet well drops 1.2 ft during the test, what is the gpm pumping rate? (Assume there is no influent to the wet well during the test.)

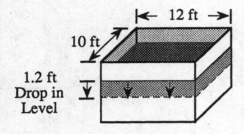

ANS_____

6. A pump must pump 900 gpm against a total head of 70 ft. What horsepower is required for this work? (Round to the nearest tenth.)

ANS_____

7. A sludge pump has a bore of 8 inches and a stroke setting of 3 inches. The pump operates at 50 revolutions per minute. If the pump operates a total of 125 minutes during a 24-hour period, what is the gpd pumping rate? (Assume the piston is 100% efficient.)

ANS_____

8. The specific gravity of a liquid is 1.4. What is the density of that liquid? (The density of water = 8.34 lbs/gal.)

ANS_____

Chapter 7—Achievement Test—Cont'd

9. What is the pressure (in psi) at a point 7 ft below the water surface?

ANS_____

10. The elevations of two water surfaces are 852 ft and 760 ft. If the friction and minor head losses equal 10 ft, what is the total dynamic head (in ft)?

ANS_____

11. What is the head loss through a gate valve (fully open) for a 6-inch diameter pipeline? (Use the nomograph provided in Appendix B.)

ANS_____

12. What is the total force exerted on the side of a tank if the tank is 15 ft wide and the water depth is 9 ft?

ANS_____

13. If 50 hp is supplied to a motor, what is the bhp and whp if the motor is 90% efficient and the pump is 85% efficient?

ANS_____

14. An 8-inch diameter pipe has a *C*-value of 100. When the flow rate is 1000 gpm, what is the friction loss for a 3000-ft length of pipe? (Use the table provided in Appendix A.)

ANS_____

15. The flow through ana 6-inch diameter pipeline is 240 gpm. The *C*-value is 100. What is the friction loss through a 1200-ft length of pipe? (Use the table provided in Appendix A.)

ANS_____

16. The motor horsepower requirement has been calculated to be 60 hp. During the week, the pump is in operation a total of 120 hours. Using a power cost of $0.5626/kWh, what would be the power cost that week for the pumping?

ANS_____

8 Water Sources and Storage

PRACTICE PROBLEMS 8.1: Well Drawdown

1. The static water level for a well is 89 ft. If the pumping water level is 97 ft, what is the well drawdown?

ANS_____

2. The static water level for a well is 105 ft. The pumping water level is 128 ft. What is the well drawdown?

ANS_____

3. Before the pump is started, the water level is measured at 142 ft. The pump is then started. If the pumping water level is determined to be 159 ft, what is the well drawdown?

ANS_____

4. The static water level of a well is 82 ft. The pumping water level is determined using the sounding line. The air pressure applied to the sounding line is 3.6 psi and the length of the sounding line is 110 ft. What is the drawdown? (Round to the nearest tenth.)

ANS_____

5. A sounding line is used to determine the static water level for a well. The air pressure applied is 4.5 psi and the length of the sounding line is 150 ft. If the pumping water level is 168 ft, what is the drawdown? (Round to the nearest tenth.)

ANS_____

PRACTICE PROBLEMS 8.2: Well Yield

1. During a 5-minute well yield test, a total of 405 gallons were pumped from the well. What is the well yield in gpm?

ANS_____

2. Once the drawdown of a well stabilized, it was determined that the well produced 780 gallons during a 5-minute pumping test. What is the well yield in gpm?

ANS_____

3. During a test for well yield, a total of 835 gallons were pumped from the well. If the well yield test lasted 5 minutes, what was the well yield in gpm? in gph?

ANS_____

ANS_____

4. A bailer is used to determine the approximate yield of a well. The bailer is 10 ft long and has a diameter of 12 inches. If the bailer is placed in the well and removed a total of 10 times during a 5-minute test, what is the well yield in gpm? (Round to the nearest tenth.)

ANS_____

5. During a 5-minute well yield test, a total of 740 gallons of water were pumped from the well. At this yield, if the pump is operated a total of 9 hours each day, how many gallons of water are pumped daily?

ANS_____

PRACTICE PROBLEMS 8.3: Specific Yield

1. The discharge capacity of a well is 190 gpm. If the drawdown is 26 ft, what is the specific yield in gpm/ft of drawdown? (Round to the nearest tenth.)

ANS_____

2. A well produces 510 gpm. If the drawdown for the well is 21 ft, what is the specific yield in gpm/ft of drawdown? (Round to the nearest tenth.)

ANS_____

3. A well yields 1000 gpm. If the drawdown is 38.2 ft, what is the specific yield in gpm/ft of drawdown? (Round to the nearest tenth.)

ANS_____

4. The specific yield of a well is listed as 31.2 gpm/ft. If the drawdown for the well is 41.6 ft, what is the well yield in gpm?

ANS_____

PRACTICE PROBLEMS 8.4: Well Casing Disinfection

1. A new well is to be disinfected with chlorine at a dosage of 50 mg/L. If the well casing diameter is 6 inches and the length of the water-filled casing is 130 ft, how many pounds of chlorine will be required? (Round to the nearest hundredth.)

ANS_____

2. A new well with a casing diameter of 12 inches is to be disinfected. The desired chlorine dosage is 50 mg/L. If the casing is 195 ft long and the water level in the well is 78 ft from the top of the well, how many pounds of chlorine will be required? (Round to the nearest tenth.)

ANS_____

3. An existing well has a total casing length of 215 ft. The top 175 ft of casing has a 12-inch diameter and the bottom 40 ft of the casing has an 8-inch diameter. The water level is 68 ft from the top of the well. How many pounds of chlorine will be required if a chlorine dosage of 100 mg/L is desired? (Round to the nearest tenth.)

ANS_____

4. The water-filled casing of a well has a volume of 520 gallons. If 0.48 lbs of chlorine were used in disinfection, what was the chlorine dosage in mg/L?

ANS_____

5. A total of 0.08 pounds of chlorine are required for the disinfection of a well. If sodium hypochlorite (5.25% available chlorine) is to be used, how many fluid ounces of sodium hypochlorite are required? (Round to the nearest tenth.)

                                            ANS_____

6. A new well is to be disinfected with calcium hypochlorite (65% available chlorine). The well casing diameter is 6 inches and the length of the water-filled casing is 115 ft. If the desired chlorine dosage is 50 mg/L, how many ounces (dry measure) of calcium hypochlorite will be required? (Round to the nearest tenth.)

                                            ANS_____

7. How many pounds of chloride of lime (25% available chlorine) will be required to disinfect a well if the casing is 18 inches in diameter and 180 ft long with a water level at 85 ft from the top of the well? The desired chlorine dosage is 100 mg/L. (Round to the nearest tenth.)

                                            ANS_____

8. The water-filled casing of a well has a volume of 235 gallons. How many fluid ounces of sodium hypochlorite (5.25% available chlorine) are required to disinfect the well if a chlorine concentration of 50 mg/L is desired?

                                            ANS_____

PRACTICE PROBLEMS 8.5: Deep-Well Turbine Pump Calculations

1. The pressure gage reading at a pump discharge head is 4.1 psi. What is this discharge head expressed in feet?

ANS_____

2. The static water level of a well is 96 ft. The well drawdown is 26 ft. If the gage reading at the pump discharge head is 3.8 psi, what is the field head?

ANS_____

3. The field head for a deep-well vertical turbine pump is 135 ft. The 10-inch diameter column is 190 ft long with a shaft enclosure tube 2 inches in diameter. If the flow through the column is 1400 gpm, what is the lab head for the pump? (Use the column friction loss table given in Appendix 1.)

ANS_____

4. The field head for a deep-well turbine pump is 175 ft. The 8-inch diameter column is 185 ft long with a shaft enclosure tube 2 inches in diameter. If the flow through the column is 950 gpm, what is the lab head for the pump? (Use the column friction loss table given in Appendix 1.)

ANS_____

5. The pumping water level for a well is 180 ft. The discharge pressure measured at the pump discharge head is 4.6 psi. If the pump capacity is 850 gpm, what is the water horsepower?

ANS_____

6. The pumping water level for a well is 205 ft. The pump discharge head is 4.8 psi. If the pump capacity is 1000 gpm, what is the water horsepower?

ANS_____

7. The bowl head of a vertical turbine pump is 178 ft and the bowl efficiency is 82%. If the capacity of the vertical turbine pump is 750 gpm, what is the bowl horsepower?

ANS_____

8. A deep-well vertical turbine pump delivers 850 gpm. The bowl head is 192 ft and the bowl efficiency is 78%. What is the bowl horsepower?

ANS_____

PRACTICE PROBLEMS 8.5—Cont'd

SHAFT FRICTION LOSS IN HP PER 100-FT OF SHAFT					
Shaft Diam. (inches)	RPM of Shaft				
	2900	1760	1450	960	860
3/4	0.51	0.31	0.26	0.17	–
1	0.87	0.53	0.44	0.29	0.26
1-1/4	1.33	0.79	0.67	0.44	0.39
1-1/2	1.90	1.14	0.96	0.63	0.56
1-3/4	2.50	1.50	1.25	0.83	0.74
2	–	1.90	1.60	1.05	0.95
1-1/4	–	2.40	2.00	1.35	1.20
Source: Floway Turbine Pumps; Peabody Floway, Inc., Fresno, CA.					

9. The bowl horsepower for a vertical turbine pump is 54.2 bhp. If the 1-1/2-inch diameter pump shaft is 170 ft long and is rotating at 1760 rpm, what is the field bhp? (Use the table given above.)

ANS_____

10. A vertical turbine pump has a bowl horsepower of 57.6 bhp. The shaft is 1-1/4 inches in diameter and rotates at a speed of 1450 rpm. If the shaft is 172 ft long, what is the field bhp?

ANS_____

11. The field brake horsepower for a deep-well turbine pump is 59.6 bhp. The thrust bearing loss is 0.5 hp. If the motor efficiency provided by the manufacturer is 90%, what is the horsepower input to the motor?

ANS_____

12. What is the bowl efficiency if the bowl head is 192 ft and the capacity is 1100 gpm? The bowl assembly has 4 stages. (Use the pump performance curve given in Appendix 2.)

ANS_____

13. What is the bowl efficiency if the bowl head is 189 ft and the capacity is 725 gpm? The bowl assemble has 3 stages. (Use the pump performance curve given in Appendix 2.)

ANS_____

14. The total brake horsepower for a deep-well turbine pump is 55.1 bhp. If the water horsepower is 44.6 whp, what is the field efficiency?

ANS_____

15. The total brake horsepower for a pump is 53.7 bhp. If the motor is 88% efficient and the water horsepower is 42.1 whp, what is the overall efficiency of the unit?

ANS_____

PRACTICE PROBLEMS 8.6: Pond or Small Lake Storage Capacity

1. A pond has an average length of 300 ft, an average width of 120 ft, and an estimated average depth of 12 ft. What is the estimated volume of the pond in gallons?

ANS_____

2. A reservoir has an average length of 350 ft and an average width of 105 ft. If the maximum depth of the reservoir is 28 ft, what is the extimated gallon volume of the lake?

ANS_____

3. A pond has an average length of 190 ft, an average width of 75 ft, and an average depth of 10 ft. What is the acre-feet volume of the pond?

ANS_____

4. A small lake has an average length of 345 ft, an average width of 185 ft, and a maximum depth of 23 ft. What is the acre-feet volume of the pond?

ANS_____

PRACTICE PROBLEMS 8.7: Copper Sulfate Dosing

1. For algae control in a lake, a dosage of 0.5 mg/L copper is desired. The lake has a volume of 25 MG. How many pounds of copper sulfate pentahydrate (25% available copper) will be required?

ANS_____

2. The desired copper dosage in a reservoir is 0.5 mg/L. The reservoir has a volume of 58 ac-ft. How many pounds of copper sulfate pentahydrate (25% available copper) will be required?

ANS_____

3. A pond has a volume of 36 ac-ft. If the desired copper sulfate dosage is 0.9 lbs CuSO$_4$/ac-ft, how many pounds of copper sulfate will be required?

ANS_____

4. A pond has an average length of 240 ft, an average width of 90 ft, and an average depth of 12 ft. If the desired dosage is 0.9 lbs copper sulfate/ac-ft, how many pounds of copper sulfate will be required?

ANS_____

5. A storage reservoir has an average length of 400 ft and an average width of 120 ft. If the desired copper sulfate dosage is 5.4 lbs $CuSO_4$/ac, how many pounds of copper sulfate will be required?

ANS_____

CHAPTER 8—ACHIEVEMENT TEST

1. The static water level for a well is 92.6 ft. If the pumping water level is 129.4 ft, what is the drawdown?

ANS_____

2. During a 5-minute well yield test, a total of 707 gallons were pumped from the well. What is the well yield in gpm? in gph?

ANS_____

ANS_____

3. A bailer is used to determine the approximate yield of a well. The bailer is 10 ft long and has a diameter of 12 inches. If the bailer is placed in the well and removed a total of 9 times during a 5-minute test, what is the well yield in gpm?

ANS_____

4. The static water level in a well is 139 ft. The pumping water level is determined using the sounding line. The air pressure applied to the sounding line is 3.7 psi and the length of the sounding line is 169 ft. What is the drawdown? (Round to the nearest tenth.)

ANS_____

5. A well produces 590 gpm. If the drawdown for the well is 26 ft, what is the specific yield in gpm/ft of drawdown? (Round to the nearest tenth.)

ANS_____

6. A new well is to be disinfected with a chlorine dose of 50 mg/L. If the well casing diameter is 6 inches and the length of the water-filled casing is 140 ft, how many pounds of chlorine will be required? (Round to the nearest hundreth.)

ANS_____

7. During a 5-minute well yield test, a total of 760 gallons of water were pumped from the well. At this yield, if the pump is operated a total of 8 hours each day, how many gallons of water are pumped daily?

ANS_____

8. The water-filled casing of a well has a volume of 590 gallons. If 0.49 lbs of chlorine were used for disinfection, what was the chlorine dosage in mg/L?

ANS_____

CHAPTER 8—ACHIEVEMENT TEST—Cont'd

9. An existing well has a total casing length of 220 ft. The top 180 ft of casing has a 12-inch diameter and the bottom 40 ft of casing has an 8-inch diameter. The water level is 72 ft from the top of the well. How many pounds of chlorine will be required if a chlorine dosage of 100 mg/*L* is desired? (Round to the nearest tenth.)

ANS_____

10. A total of 0.1 lbs of chlorine are required for the disinfection of a well. If sodium hypochlorite is to be used (5.25% available chlorine), how many fluid ounces of sodium hypochlorite are required? (Round to the nearest tenth.)

ANS_____

11. The pressure gage reading at a pump discharge head is 4.3 psi. What is this discharge head expressed in feet?

ANS_____

12. The static water level of a well is 98 ft. The well drawdown is 27 ft. If the gage reading at the pump discharge head is 3.9 psi, what is the field head?

ANS_____

13. The field head for a deep-well turbine pump is 180 ft. The 8-inch diameter column is 190 ft long with a shaft enclosure tube 2 inches in diameter. If the flow through the column is 1200 gpm, what is the lab head for the pump? (Use the column friction loss table given in Appendix 1.)

ANS_____

14. The pumping water level for a well is 187 ft. The discharge pressure measured at the pump discharge head is 4.3 psi. If the pump capacity is 900 gpm, what is the horsepower?

ANS_____

15. A deep-well vertical turbine pump delivers 825 gpm. The bowl head is 178 ft and the bowl efficiency is 80%. What is the bowl horsepower?

ANS_____

CHAPTER 8—ACHIEVEMENT TEST—Cont'd

SHAFT FRICTION LOSS IN HP PER 100-FT OF SHAFT					
Shaft Diam. (inches)	RPM of Shaft				
	2900	1760	1450	960	860
3/4	0.51	0.31	0.26	0.17	–
1	0.87	0.53	0.44	0.29	0.26
1-1/4	1.33	0.79	0.67	0.44	0.39
1-1/2	1.90	1.14	0.96	0.63	0.56
1-3/4	2.50	1.50	1.25	0.83	0.74
2	–	1.90	1.60	1.05	0.95
1-1/4	–	2.40	2.00	1.35	1.20

Source: Floway Turbine Pumps; Peabody Floway, Inc., Fresno, CA.

16. The bowl horsepower for a vertical turbine pump is 56.4 bhp. If the 1-1/2-inch diameter pump shaft is 174 ft long and is rotating at 1760 rpm, what is the field bhp? (Use the table given above.)

ANS_____

17. The field brake horsepower for a deep-well turbine pump is 49.6 bhp. The thrust bearing loss is 0.6 hp. If the motor efficiency provided by the manufacturer is 90%, what is the horsepower input to the motor?

ANS_____

18. What is the bowl efficiency if the bowl head is 192 ft and the capacity is 675 gpm? The bowl assembly has 3 stages. (Use the pump performance curve given in Appendix 2.)

ANS_____

19. The total bhp for a deep-well turbine pump is 56.2 bhp. If the water horsepower is 44.8 whp, what is the field efficiency?

ANS_____

20. The total brake horsepower for a pump is 53.9 bhp. If the motor is 90% efficient and the water horsepower is 43.5 whp, what is the overall efficiency of the unit?

ANS_____

21. The desired copper dosage at a reservoir is 0.5 mg/*L*. The reservoir has a volume of 51 ac-ft. How many pounds of copper sulfate pentahydrate (25% available copper will be required?

ANS_____

22. A storage reservoir has an average length of 420 ft and an average width of 130 ft. If the desired copper sulfate dosage is 5.4 lbs copper sulfate/ac, how many pounds of copper sulfate will be required?

ANS_____

9 Coagulation and Flocculation

PRACTICE PROBLEMS 9.1: Chamber or Basin Volume

1. A flash mix chamber is 3 ft wide, 4 ft long, with water to a depth of 3 ft. What is the gallon volume of water in the flash mix chamber?

ANS_____.

2. A flocculation basin is 45 ft long, 15 ft wide, with water to a depth of 9 ft. What is the volume of water in the basin (in gallons)?

ANS_____

3. A flocculation basin is 35 ft long, 15 ft wide, with water to a depth of 8 ft. How many gallons of water are in the basin?

ANS_____

4. A flash mix chamber is 4 ft square with water to a depth of 40 inches. What is the volume of water in the flash mixing chamber (in gallons)?

ANS_____

5. A flocculation basin is 20 ft wide, 35 ft long and contains water to a depth of 8 ft 10 in.. What is the volume of water (in gallons) in the flocculation basin?

ANS_____

PRACTICE PROBLEMS 9.2: Detention Time

1. The flow to a flocculation basin is 3,540,000 gpd. If the basin is 50 ft long, 20 ft wide with water to a depth of 8 ft, what is the detention time (in minutes) of the flocculation basin?

ANS_____

2. A flocculation basin is 45 ft long, 15 ft wide, and has a water level of 9 ft. What is the detention time (in minutes) in the basin if the flow to the basin is 2.6 MGD?

ANS_____

3. A flash mix chamber 5 ft long, 4 ft wide and 4.5 ft deep receives a flow of 8 MGD. What is the detention time in the chamber (in seconds)?

ANS_____

4. A flocculation basin is 40 ft long, 15 ft wide, and has a water depth of 8 ft 8 inches. If the flow to the basin is 2,100,000 gpd, what is the detention time in minutes?

ANS_____

5. A flash mix chamber is 3 ft square, with a water depth of 40 inches. If the flash mix chamber receives a flow of 3.15 MGD, what is the detention time in seconds?

ANS_____

**PRACTICE PROBLEMS 9.3: Determining Chemical Feeder Setting—
Dry Chemical Feeder, lbs/day**

1. The desired dry alum dosage, as determined by the jar test, is 9 mg/L. Determine the lbs/day setting on a dry alum feeder if the flow is 3,410,000 gpd.

ANS_____

2. Jar tests indicate that the best polymer dose for a water is 13 mg/L. If the flow to be treated is 1,726,000 gpd, what should the dry chemical feed setting be, in lbs/day?

ANS_____

3. Determine the desired lbs/day setting on a dry alum feeder if jar tests indicate an optimum dose of 11 mg/L and the flow to be treated is 2.82 MGD.

ANS_____

4. The desired dry alum dose is 8 mg/*L*, as determined by a jar test. If the flow to be treated is 960,000 gpd, how many lbs/day dry alum will be required?

ANS_____

5. A flow of 4.05 MGD is to be treated with a dry polymer. If the desired dose is 14 mg/*L* what should the dry chemical feeder setting be (lbs/day)?

ANS_____

PRACTICE PROBLEMS 9.4: Determining Chemical Feeder Setting— Solution Chemical Feeder, gpd

1. Jar tests indicate that the best alum dose for a water is 8 mg/L. The flow to be treated is 1.82 MGD. Determine the gpd setting for the alum solution feeder if the liquid alum contains 5.36 lbs of alum per gallon of solution.

ANS_____

2. The flow to a plant is 3.12 MGD. Jar testing indicates that the optimum alum dose is 13 mg/L. What should the gpd setting be for the soution feeder if the alum solution is a 55% solution?

ANS_____

3. Jar tests indicate that the best alum dose for a water is 11 mg/L. The flow to be treated is 4.02 MGD. Determine the gpd setting for the alum solution feeder if the liquid alum contains 5.45 lbs of alum per gallon of solution.

ANS_____

4. Jar tests indicate that the best liquid alum dose for a water is 10 mg/*L*. The flow to be treated is 940,000 gpd. Determine the gpd setting for the liquid alum chemical feeder if the liquid alum is a 60% solution.

ANS_____

5. A flow of 1,675,000 gpd is to be treated with alum. Jar test indicate that the optimum alum dose is 9 mg/*L*. If the liquid alum contains 642 mg alum/m*L* solution, what should be the gpd setting for the alum solution feeder?

ANS_____

PRACTICE PROBLEMS 9.5: Determining Chemical Feeder Setting— Solution Chemical Feeder, m*L*/min

1. The desired solution feed rate was calculated to be 38 gpd. What is this feed rate expressed as m*L*/min?

ANS_____

2. The desired solution feed rate was calculated to be 24.7 gpd. What is this feed rate expressed as m*L*/min?

ANS_____

3. The optimum polymer dose has been determined to be 12 mg/*L*. The flow to be treated is 2,920,000 gpd. If the solution to be used contains 60% active polymer, what should the solution chemical feeder setting be, in m*L*/min?

ANS_____

4. The optimum polymer dose for a 2,670,000 gpd-flow has been determined to be 8 mg/*L*. If the polymer solution contains 58% active polymer, what should the solution chemical feeder setting be, in m*L*/min? (Assume the polymer solution weighs 8.34 lbs/gal.)

ANS_____

5. Jar tests indicate that the best alum dose for a water is 11 mg/*L*. The liquid alum contains 5.42 lbs alum per gallon of solution. What should the setting be on the solution chemical feeder (in m*L*/min) when the flow to be treated is 3.61 MGD?

ANS_____

PRACTICE PROBLEMS 9.6: Percent Strength of Solutions

1. If 130 grams of dry polymer are dissolved in 15 gallons of water, what is the percent strength of the solution ? (1 g = 0.0022 lbs) (Round lbs polymer to the nearest tenth.)

ANS_____

2. If a total of 20 ounces of dry polymer are added to 20 gallons of water, what is the percent strength (by weight) of the polymer solution. (Round the pounds polymer to the nearest tenth when making this calculation.)

ANS_____

3. How many gallons of water must be added to 1.9 lbs dry alum to make a 0.9% solution?

ANS_____

4. A 10% liquid polymer is to be used in making up a polymer solution. How many lbs of liquid polymer should be mixed with water to produce 170 lbs of a 0.6% polymer solution?

ANS_____

5. A 9% polymer solution will to be used in making up a solution. How many gallons of liquid polymer should be added to the water to make up 40 gallons of a 0.3% polymer solution? The liquid polymer has a specific gravity of 1.3. Assume the polymer solution has a specific gravity of 1.0.

ANS_____

6. How many gallons of an 11% liquid polymer should be mixed with water to produce 75 gallons of a 0.7% polymer solution? The Density of the polymer liquid is 10.1 lbs/gal. Assume the density of the polymer solution is 8.34 lbs/gal.

ANS_____

PRACTICE PROBLEMS 9.7: Mixing Solutions of Different Strength

1. If 30 lbs of a 10% strength solution are mixed with 70 lbs of a 0.5% strength solution, what is the percent strength of the solution mixture? (Round to the nearest tenth.)

ANS_____

2. If 5 gallons of a 15% strength solution are added to 45 gallons of a 0.25% strength solution, what is the percent strength of the solution mixture? Assume the 15% strength solution weighs 10.9 lbs/gal and the 0.25% strength solution weighs 8.38 lbs/gal. (Round to the nearest tenth.)

ANS_____

3. If 10 gallons of a 12% strength solution is mixed with 40 gallons of a 0.75% strength solution, what is the percent strength of the solution mixture? Assume the 12% solution weighs 10.5 lbs/gal and the 0.75% solution weighs 8.42 lbs/gal. (Round to the nearest tenth.)

ANS_____

4. What weights of a 2% and 7% solution must be mixed to make 400 lbs of a 4% solution?

ANS_____

5. How many lbs of a 1% polymer solution and water should be mixed together to form 75 lbs of a 0.7% solution?

ANS_____

6. What weights of a 0.5% solution and an 8% solution must be mixed to make 200 lbs of a 3% solution?

ANS_____

PRACTICE PROBLEMS 9.8: Dry Chemical Feeder Calibration

1. Calculate the actual chemical feed rate, in lbs/day, if a bucket is placed under a chemical feeder and a total of 2.1 lbs is collected during a 30-minute period.

ANS_____

2. Calculate the actual chemical feed rate, in lbs/day, if a bucket is placed under a chemical feeder and a total of 40 oz is collected during a 45-minute period.

ANS_____

3. To calibrate a chemical feeder, a bucket is first weighed (12 oz) then placed under the chemical feeder. Ater 30 minutes the bucket is weighed again. If the weight of the bucket with chemical is 2.2 lbs, what is the actual chemical feed rate, in lbs/day?

ANS_____

4. A chemical feeder is to be calibrated. The bucket to be used to collect chemical is placed under the chemical feeder and weighed (0.5 lbs). After 30 minutes, the weight of the bucket and chemical is found to be 2.7 lbs. Based on this test, what is the actual chemical feed rate, in lbs/day?

ANS_____

5. During a 24-hour period, a flow of 1,806,000 gpd water is treated. If a total of 35 lbs of polymer were used for coagulation during that 24-hour period, what is the polymer dosage in mg/*L*?

ANS_____

PRACTICE PROBLEMS 9.9: Solution Chemical Feeder Calibration
(Given Flow Rate)

1. A calibration test is conducted for a solution chemical feeder. During a 24-hour period a total of 75 gallons solution is delivered by the solution feeder. The polymer solution is a 1.5% solution. What is the lbs/day solution feed rate? Assume the polymer solution weighs 8.34 lbs/gal. (Round to the nearest tenth.)

ANS_____

2. A calibration test is conducted for a solution chemical feeder. During a 5-minute test, the pump delivered 620 mL of a 1.3% polymer solution. The specific gravity of the polymer solution is 1.09. What is the polymer dosage rate in lbs/day? (Round to the nearest tenth.)

ANS_____

3. During a 5-minute calibration test for a solution chemical feeder, a total of 710 mL is delivered by the solution feeder. The polymer solution is a 1.1% solution. What is the lbs/day polymer feed rate? (Assume the polymer solution weighs 8.34 lbs/gal.)

ANS_____

4. A solution chemical feeder delivered 930 m*L* solution during a 5-minute calibration test. The polymer solution is a 1.35% strength solution. What is the polymer dosage rate in lbs/day? Assume the polymer solution weighs 8.34 lbs/gal. (Round to the nearest tenth.)

ANS_____

5. If 1680 m*L* of a 1.8% polymer solution are delivered during a 10-minute calibration test, and the polymer solution has a specific gravity of 1.09, what is the polymer feed rate in lbs/day? (Round to the nearest tenth.)

ANS_____

PRACTICE PROBLEMS 9.10: Solution Chemical Feeder Calibration
(Given Drop in Solution Tank Level)

1. A pumping rate calibration test is conducted for a 5-minute period. The liquid level in the 3-ft diameter solution tank is measured before and after the test. If the level drops 4 inches during the 5-minute test, what is the gpm pumping rate?

ANS_____

2. During a 10-minute pumping rate calibration test, the liquid level in the 3-ft diameter solution tank drops 3 inches. What is the pumping rate in gpm?

ANS_____

3. The liquid level in a 3-ft diameter solution tank drops 2 inches during a 10-minute pumping rate calibration test. What is the gpm pumping rate? Assuming a continuous pumping rate, what is the gpd pumping rate?

ANS_____

4. During a 20-minute pumping rate calibration test, the solution level in the 3-ft diameter chemical tank dropped 1.5 inches. Assume the pump operates at the rate during the next 24 hours. If the polymer solution is a 1.2% solution, what is the lbs/day polymer feed? (Assume the polymer solution weighs 8.34 lbs/gal.)

ANS_____

5. The level in a 3-ft diameter chemical tank drops 1 inch during a 30-minute pumping rate calibration test. The polymer solution is a 1.25% solution. If the pump operates at the same rate for 24 hours, what is the polymer feed in lbs/day?

ANS_____

PRACTICE PROBLEMS 9.11: Chemical Use Calculations

1. The amount of chemical used for each day during a week is given below. Based on this data, what was the average lbs/day chemical use during the week?

Monday—78 lbs/day	Friday—77 lbs/day
Tuesday—71 lbs/day	Saturday—82 lbs/day
Wednesday—72 lbs/day	Sunday—85 lbs/day
Thursday—65 lbs/day	

    ANS_____

2. The average chemical use at a plant is 95 lbs/day. If the chemical inventory in stock is 2100 lbs, how many day's supply is this? (Round to the nearest tenth.)

    ANS_____

3. The chemical inventory in stock is 921 lbs. If the average chemical use at a plant is 60 lbs/day, how many day's supply is this? (Round to the nearest tenth.)

    ANS_____

4. The average gallons of polymer solution used each day at a treatment plant is 92 gpd. A chemical feed tank has a diameter of 4 ft and contains solution to a depth of 3 ft 2 inches. How many day's supply are represented by the solution in the tank? (Round to the nearest tenth.)

ANS_____

5. Jar tests indicate that the optimum polymer dose for a water is 2.5 mg/*L*. If the flow to be treated is 1.6 MGD, how many lbs of dry polymer will be required for a 30-day period?

ANS_____

Chapter 9—Achievement Test

1. A flash mix chamber 4 ft long, 3 ft wide, with a 3-ft water depth receives a flow of 5.7 MGD. What is the detention time in the chamber (in seconds)?

ANS_____

2. A flocculation basin is 40 ft long, 15 ft wide, and has a water depth of 8 ft. What is the volume of water in the basin (in gallons)?

ANS_____

3. The desired dry alum dosage, as determined by jar testing, is 7 mg/*L*. Determine the lbs/day setting on a dry alum feeder if the flow is 3.62 MGD.

ANS_____

4. The flow to a plant is 2.95 MGD. Jar testing indicates that the best alum dose is 9 mg/*L*. What should the gpd setting be for the solution feeder if the alum solution is a 55% solution? (Assume the alum solution weighs 8.34 lbs/gal.)

ANS_____

5. A flash mix chamber is 4 ft square with a water depth of 3 ft. What is the gallon volume of water in this chamber?

ANS_____

6. The desired solution feed rate was calculated to be 40 gpd. What is this feed rate expressed as mL/min?

ANS_____

7. A flocculation basin is 45 ft long, 20 ft wide, and has a water depth of 9 ft 9 in. If the flow to the basin is 2,480,000 gpd, what is the detention time in minutes?

ANS_____

8. The optimum polymer dose has been determined to be 7 mg/L. The flow to be treated is 1,960,000 gpd. If the solution to be used contains 60% active polymer, what should the solution chemical feeder setting be, in mL/min? (The polymer solution weighs 10.2 lbs/gal.)

ANS_____

Chapter 9—Achievement Test—Cont'd

9. The desired solution feed rate was calculated to be 85 gpd. What is this feed rate expressed as mL/min?

ANS_____

10. Determine the desired lbs/day setting on a dry alum feeder if jar tests indicate an optimum dose of 8 mg/L and the flow to be treated is 940,000 gpd.

ANS_____

11. How many gallons of water must be added to 2.5 lbs of dry alum to make a 1.3% solution?

ANS_____

12. If 20 lbs of an 15% strength solution are mixed with 130 lbs of a 0.6% strength solution, what is the percent strength of the solution mixture?

ANS_____

13. Calculate the chemical feed rate, in lbs/day , if a bucket is placed under a chemical feeder and a total of 3.5 lbs chemical is collected during a 30-minute period.

ANS_____

14. A chemical feeder is to be calibrated. The bucket to be used to collect chemical is placed under the chemical feeder and weighed (1.5 lbs). After 30 minutes, the weight of the bucket and chemical is found to be 3.9 lbs. Based on this test, what is the chemical feed rate, in lbs/day?

ANS_____

15. If 180 grams of dry polymer are dissolved in 20 gallons of water, what is the percent strength (by weight) of the solution? (1 g = 0.0022 lbs)

ANS_____

16. During a 5-minute calibration test for a solution chemical feeder, a total of 750 m*L* is delivered by the feeder. The polymer solution is a 1.5% solution. What is the lbs/day polymer feed rate? (Assume the polymer solution weighs 8.34 lbs/gal.)

ANS_____

Chapter 9—Achievement Test—Cont'd

17. Jar tests indicate that the best alum dose for a water is 12 mg/*L*. The flow to be treated is 4.1 MGD. Determine the gpd setting for the alum solution feeder if the liquid alum contains 5.36 lbs of alum per gallon of solution.

ANS_____

18. A 10% liquid polymer is to be used in making up a polymer solution. How many lbs of liquid polymer should be mixed with water to produce 200 lbs of a 0.7% polymer solution?

ANS_____

19. How many lbs of a 55% polymer solution and water should be mixed together to form 150 lbs of a 1% solution?

ANS_____

20. During a 10-minute pumping rate calibration test, the liquid level in the 3-ft diameter solution tank drops 2 inches. What is the pumping rate in gpm?

ANS_____

21. How many gallons of a 10% liquid polymer should be mixed with water to produce 75 gallons of a 0.5% polymer solution? The density of the polymer liquid is 9.8 lbs/gal. Assume the density of the polymer solution is 8.34 lbs/gal.

ANS_____

22. A calibration test is conducted of a solution chemical feeder. During a 5-minute test, the pump delivered 680 m*L* of a 0.9% polymer solution. The specific gravity of the polymer solution is 1.1. What is the polymer dosage rate in lbs/day?

ANS_____

23. Jar tests indicate that the best polymer dose for a water is 5 mg/*L*. If the flow to be treated is 3.4 MGD, at this rate how many lbs of dry polymer will be required for a 30-day period?

ANS_____

24. The chemical inventory in stock is 530 lbs. If the average chemical use at the plant is 70 lbs/day, how many days' supply is this?

ANS_____

10 *Sedimentation*

PRACTICE PROBLEMS 10.1: Tank Volume

1. A sedimentation basin is 60 ft long and 25 ft wide. If the water depth is 12 ft, what is the volume of water in the tank, in gallons?

ANS_____

2. A circular clarifier has a diameter of 70 ft. If the water depth is 10 ft, how many gallons of water are in the tanks?

ANS_____

3. A sedimentation tank is 75 ft long, 25 ft wide, and has water to a depth of 12 ft. What is the volume of water in the tank, in gallons?

ANS_____

4. A sedimentation basin is 50 ft long and 20 ft wide. When the basin contains a total of 60,000 gallons, what would be the water depth?

ANS_____

5. A circular clarifier is 70 ft in diameter. If the water depth is 10 ft 3 inches, what is the volume of water in the clarifier, in gallons?

ANS_____

PRACTICE PROBLEMS 10.2: Detention Time

1. A rectangular sedimentation basin is 80 ft long, 30 ft wide, and has water to a depth of 12 ft. The flow to the basin is 2,120,000 gpd. Calculate the detention time in hours for the sedimentation basin.

ANS_____

2. A circular clarifier has a diameter of 70 ft and an average water depth of 11 ft. If the flow to the clarifier is 2,815,000 gpd, what is the detention time, in hours?

ANS_____

3. A rectangular sedimentation basin is 50 ft long and 15 ft wide, with water to a depth of 12 ft. If the flow to the basin is 1,470,000 gpd, what is the detention time, in hours?

ANS_____

4. A circular clarifier has a diameter of 50 ft and an average water depth of 10 ft. What flow rate (MGD) corresponds to a detention time of 2.5 hrs?

ANS_____

5. A sedimentation basin is 60 ft long and 20 ft wide. The average water depth is 10 ft. If the flow to the basin is 1,570,000 gpd, what is the detention time in the sedimentation basin in hours?

ANS_____

PRACTICE PROBLEMS 10.3: Surface Overflow Rate

1. A rectangular sedimentation basin is 50 ft long and 20 ft wide. When the flow is 490 gpm, what is the surface overflow rate in gpm/sq ft?

ANS_____

2. A circular clarifier has a diameter of 60 ft. If the flow to the clarifier is 1550 gpm, what is the surface overflow rate in gpm/sq ft?

ANS_____

3. A rectangular sedimentation basin receives a flow of 530,000 gpd. If the basin is 45 ft long and 20 ft wide, what is the surface overflow rate in gpm/sq ft?

ANS_____

4. A sedimentation basin is 75 ft long and 25 ft wide. To maintain a surface overflow rate of 0.5 gpm/sq ft, what is the maximum flow to the basin in gpd?

ANS_____

5. A circular clarifier 50 ft in diameter receives a flow of 1,670,000 gpd. What is the surface overflow rate in gpm/sq ft?

ANS_____

PRACTICE PROBLEMS 10.4: Mean Flow Velocity

1. A sedimentation basin is 70 ft long, 20 ft wide, and operates at a depth of 12 ft. If the flow to the basin is 1,440,000 gpd, what is the mean flow velocity, in fpm?

ANS_____

2. A sedimentation basin is 60 ft long, 25 ft wide, and operates at a depth of 10 ft. If the flow to the basin is 1.6 MGD, what is the mean flow velocity, in fpm?

ANS_____

3. A sedimentation basin is 75 ft long and 30 ft wide. The water level is 13 ft. When the flow to the basin is 2.24 MGD, what is the mean flow velocity, in fpm?

ANS_____

4. The flow to a sedimentation basin is 2,950,000 gpd. If the length of the basin is 80 ft, the width of the basin is 30 ft, and the depth of water in the basin is 12 ft, what is the mean flow velocity in fpm?

ANS_____

5. A sedimentation basin 40 ft long and 20 ft wide receives a flow of 890,000 gpd. The basin operates at a depth of 10 ft. What is the mean flow velocity in the basin, in ft/min?

ANS_____

PRACTICE PROBLEMS 10.5: Weir Loading Rate

1. A circular clarifier receives a flow of 2,420,000 gpd. If the diameter of the weir is 60 ft, what is the weir loading rate in gpm/ft?

ANS_____

2. A rectangular sedimentation basin has a total of 170 ft of weir. If the flow to the basin is 1,940,000 gpd, what is the weir loading rate in gpm/ft?

ANS_____

3. A rectangular sedimentation basin has a total of 110 ft of weir. If the flow over the weir is 1,260,000 gpd, what is the weir loading rate in gpm/ft?

ANS_____

4. A circular clarifier receives a flow of 3.1 MGD. If the diameter of the weir is 75 feet, what is the weir loading rate in gpm/ft?

ANS_____

5. A rectangular sedimentation basin has a total of 150 ft of weir. If the flow over the weirs is 1.8 MGD, what is the weir loading rate in gpm/ft?

ANS_____

PRACTICE PROBLEMS 10.6: Percent Settled Sludge
(V/V Test)

1. A 100-mL sample of slurry from a solids contact unit is placed in a graduated cylinder and allowed to settle for 10 minutes. The settled sludge at the bottom of the graduated cylinder after 10 minutes is 21 mL. What is the percent settled sludge of the sample?

ANS_____

2. A 100-mL sample of slurry from a solids contact unit is placed in a graduated cylinder. After 10 minutes, a total of 23 mL of sludge settled to the bottom of the cylinder. What is the percent settled sludge of the sample?

ANS_____

3. A percent settled sludge test is conducted on a 100-mL sample of solids contact unit slurry. After 10 minutes of settling, a total of 17-mL of sludge is found to settle to the bottom of the cylinder. What is the percent settled sludge of the sample?

ANS_____

4. A 100-mL sample of slurry from a solids contact unit is placed in a graduated
cylinder. After 10 minutes, a total of 18 mL of sludge is found to have settled to the
bottom of the cylinder. What is the percent settled sludge of the sample?

ANS_____

PRACTICE PROBLEMS 10.7: Lime Dose Required, mg/L

1. A raw water requires an alum dose of 48 mg/L, as determined by jar testing. If a "residual" 30 mg/L alkalinity (HCO$_3^-$) must be present in the water to promote precipitation of the alum added, what is the total alkalinity required, in mg/L? (1 mg/L alum reacts with 0.45 mg/L alkalinity, HCO$_3^-$.)

ANS_____

2. Jar tests indicate that 50 mg/L alum are optimum for a particular raw water. If a "residual" 30 mg/L alkalinity must be present to promote complete precipitation of the alum added, what is the total alkalinity required, in mg/L? (1 mg/L alum reacts with 0.45 mg/L alkalinity, HCO$_3^-$.)

ANS_____

3. A total of 45 mg/L alkalinity (HCO$_3^-$) is required to react with alum and ensure proper precipitation. If the raw water has an alkalinity of 28 mg/L as bicarbonate (HCO$_3^-$), how many mg/L alkalinity (HCO$_3^-$) should be added to the water?

ANS_____

4. A total of 42 mg/L alkalinity is required to react with alum and ensure complete precipitation of the alum added. If the raw water has an alkalinity of 26 mg/L as bicarbonate (HCO$_3^-$), how many mg/L alkalinity (HCO$_3^-$) should be added to the water?

ANS_____

5. A total of 17 mg/*L* alkalinity (HCO_3^-) must be added to a raw water. How many mg/*L* lime will be required to provide this amount of alkalinity? (1 mg/*L* alum reacts with 0.45 mg/*L* alkalinity and 1 mg/*L* alum reacts with 0.35 mg/*L* lime.)

ANS_____

6. It has been calculated that 21 mg/*L* alkalinity (HCO_3^-) must be added to a raw water. How many mg/*L* lime will be required to provide this amount of alkalinity? (1 mg/*L* alum reacts with 0.45 mg/*L* alkalinity and 1 mg/*L* alum reacts with 0.35 mg/*L* lime.)

ANS_____

7. Given the following data, calculate the required lime dose, in mg/*L*.

- Alum dose required per jar tests—52 mg/*L*

- Raw water alkalinity—36 mg/*L*

- "Residual" alkalinity req'd for precipitation—30 mg/*L*

- 1 mg/*L* alum reacts with 0.45 mg/*L* alkalinity (HCO_3^-)

- 1 mg/*L* alkalinity reacts with 0.35 mg/*L* lime

ANS_____

PRACTICE PROBLEMS 10.8: Lime Dose Required, lbs/day

1. The lime dose for a raw water has been calculated to be 13.6 mg/*L*. If the flow to be treated is 2.4 MGD, how many lbs/day lime will be required?

ANS_____

2. The lime dose for a solids contact unit has been calculated to be 12.9 mg/*L*. If the flow to be treated is 2,150,000 gpd, how many lbs/day lime will be required?

ANS_____

3. The flow to a solids contact clarifier is 970,000 gpd. If the lime dose required is determined to be 15.2 mg/*L*, how many lbs/day lime will be required?

ANS_____

4. A solids contact clarification unit receives a flow of 1.1 MGD. Alum is to be used for coagulation purposes. If a lime dose of 14 mg/*L* will be required, how many lbs/day lime is this?

ANS_____

PRACTICE PROBLEMS 10.9: Lime Dose Required, g/min

1. A total of 195 lbs/day lime will be required to raise the alkalinity of the water passing through a solids-contact clarifier. How many g/min lime does this represent?
(1 lb = 453.6 g)

ANS_____

2. The lime dose of 105 lbs/day is required for a raw water passing through a solids-contact clarifier. How many g/min lime does this represent? (1 lb = 453.6 g)

ANS_____

3. A lime dose of 14 mg/L is required to raise the pH of a particular water. If the flow to be treated is 810,000 gpd, what g/min lime dose will be required? (1 lb = 453.6 g)

ANS_____

4. A lime dose of 12.5 mg/*L* is required to raise the alkalinity of a raw water. If the flow to be treated is 2,840,000 gpd, what is the g/min lime dose required? (1 lb = 453.6 g)

ANS_____

Chapter 10—Achievement Test

1. A rectangular sedimentation basin is 55 ft long and 20 ft wide, with water to a depth of 10 ft. If the flow to the basin is 1,270,000 gpd, what is the detention time, in hours?

    ANS_____

2. A sedimentation basin is 65 ft long and 25 ft wide. If the water depth is 12 ft, what is the volume of water in the tank, in gallons?

    ANS_____

3. A sedimentation basin 70 ft long and 20 ft wide operates at a depth of 11 ft. If the flow to the basin is 1,510,000 gpd, what is the mean flow velocity, in fpm?

    ANS_____

4. A rectangular sedimentation basin receives a flow of 615,000 gpd. If the basin is 45 ft long and 20 ft wide, what is the surface overflow rate in gpm/sq ft?

    ANS_____

5. A circular clarifier has a diameter of 60 ft. If the water depth is 12 ft, how many gallons of water are in the tank?

ANS_____

6. A rectangular sedimentation basin has a total of 175 ft of weir. If the flow to the basin is 2,110,000 gpd, what is the weir loading rate in gpm/ft?

ANS_____

7. A circular clarifier has a diameter of 70 ft and an average water depth of 12 ft. If the flow to the clarifier is 2.92 MGD, what is the detention time, in hours?

ANS_____

8. A sedimentation basin 60 ft long and 25 ft wide operates at a depth of 10 ft. If the flow to the basin is 1.72 MGD, what is the mean flow velocity, in fpm?

ANS_____

Chapter 10—Achievement Test

9. A circular clarifier has a diameter of 60 ft. If the flow to the clarifier is 1600 gpm, what is the surface overflow rate in gpm/sq ft?

ANS_____

10. A circular clarifier receives a flow of 2.51 MGD. If the diameter of the weir is 60 ft, what is the weir loading rate in gpm/ft?

ANS_____

11. A circular clarifier has a diameter of 50 ft and an average water depth of 11 ft. What flow rate (MGD) corresponds to a detention time of 2 hours?

ANS_____

12. The flow to a sedimentation basin is 3.12 MGD. If the length of the basin is 75 ft, the width of the basin is 30 ft, and the depth of water in the basin is 12 ft, what is the mean flow velocity, in ft/min?

ANS_____

13. A 100-m*L* sample of slurry from a solids contact clarification unit is placed in a graduated cylinder and allowed to settle for 10 minutes. The settled sludge at the bottom of the graduated cylinder after 10 minutes is 24 m*L*. What is the percent settled sludge of the sample?

ANS_____

14. A raw water requires an alum dose of 50 mg/*L*, as determined by jar testing. If a "residual" 30 mg/*L* alkalinity (HCO_3^-) must be present in the water to promote complete precipitation of the alum added, what is the total alkalinity required, in mg/*L*? (1 mg/*L* alum reacts with 0.45 mg/*L* alkalinity, HCO_3^-.)

ANS_____

15. A sedimentation basin is 75 ft long and 25 ft wide. To maintain a surface overflow rate of 0.6 gpm/sq ft, what is the maximum flow to the basin in gpd?

ANS_____

16. The lime dose for a raw water has been calculated to be 14.2 mg/*L*. If the flow to be treated is 2,370,000 gpd, how many lbs/day lime will be required?

ANS_____

Chapter 10—Achievement Test—Cont'd

17. A 100-m*L* sample of slurry from a solids contact clarification unit is placed in a graduated cylinder. After 10 minutes, a total of 19 m*L* of sludge has settled to the bottom of the cylinder. What is the percent settled sludge of the sample?

ANS_____

18. A circular clarifier receives a flow of 3.13 MGD. If the diameter of the weir is 75 ft, what is the weir loading rate in gpm/ft?

ANS_____

19. A total of 40 mg/*L* alkalinity (HCO_3^-) is required to react with alum and ensure complete precipitation of the alum added. If the raw water has an alkalinity of 28 mg/*L* as bicarbonate (HCO_3^-), how many mg/*L* alkalinity (HCO_3^-) should be added to the water?

ANS_____

20. Given the following data, calculate the required lime dose, in mg/*L*.

- Alum dose required per jar tests—48 mg/*L*

- Raw water alkalinity—32 mg/*L*

- "Residual" alkalinity req'd for precipitation—30 mg/*L*

- 1 mg/*L* alum reacts with 0.45 mg/*L* alkalinity (HCO_3^-)

- 1 mg/*L* alkalinity reacts with 0.35 mg/*L* lime

ANS_____

21. A total of 188 lbs/day lime will be required to raise the alkalinity of the water passing through a solids contact clarifier. How many grams per minute lime does this represent? (1 lb = 453.6 grams)

ANS_____

22. Given the data below, calculate the required lime dose, in mg/L

- Alum dose required per jar tests—50 mg/L

- Raw water alkalinity—34 mg/L

- "Residual" alkalinity req'd for precipitation—30 mg/L

- 1 mg/L alum reacts with 0.45 mg/L alkalinity (HCO_3^-)

- 1 mg/L alkalinity reacts with 0.35 mg/L lime

ANS_____

23. A solids contact clarification unit receives a flow of 1.3 MGD. Alum is to be used for coagulation purposes. If a lime dose of 15 mg/L is required, how many lbs/day lime is this?

ANS_____

24. A lime dose of 13 mg/L is required to raise the alkalinity of a raw water. If the flow to be treated is 2,940,000 gpd, what is the g/min lime dose required?

ANS_____

11 *Filtration*

PRACTICE PROBLEMS 11.1: Flow Rate Through A Filter

1. During an 80-hour filter run, a total of 13.5 million gallons of water are filtered. What is the average gpm flow rate through the filter during this time?

ANS_____

2. The flow rate through a filter is 2.77 MGD. What is this flow rate expressed as gpm?

ANS_____

3. At an average flow rate through a filter of 3000 gpm, how long a filter run (in hours) would be required to produce 14 MG of filtered water?

ANS_____

4. The influent valve to a filter is closed for a 5-minute period. During this time, the water level in the filter drops 13 inches. If the filter is 40 ft long and 25 ft wide, what is the gpm flow rate through the filter?

ANS_____

5. A filter is 30 ft long and 25 ft wide. To verify the flow rate through the filter, the filter influent valve is closed for a 5-minute period and the water drop is measured. If the water level in the filter drops 16 inches during the 5 minutes, what is the gpm flow rate through the filter?

ANS_____

6. The influent valve to a filter is closed for 8 minutes. The water level in the filter drops 20 inches during the 8 minutes. If the filter is 40 ft long and 20 ft wide, what is the gpm flow rate through the filter?

ANS_____

PRACTICE PROBLEMS 11.2: Filtration Rate, gpm/sq ft

1. A filter 25 ft long and 20 ft wide receives a flow of 1930 gpm. What is the filtration rate in gpm/sq ft?

ANS_____

2. A filter has a surface area of 35 ft by 20 ft. If the filter receives a flow of 3,360,000 gpd, what is the filtration rate in gpm/sq ft?

ANS_____

3. A filter 40 ft long and 25 ft wide produces a total of 18.7 MG during a 73.4-hr filter run. What is the average filtration rate for this filter run?

ANS_____

4. A filter 35 ft long and 25 ft wide produces a total of 14.9 MG during a 72.6-hr filter run. What is the average filtration rate for this filter run?

ANS_____

5. A filter 40 ft long and 25 ft wide receives a flow of 3,740,000 gpd. What is the filtration rate in gpm/sq ft?

ANS_____

6. A filter is 40 ft long and 20 ft wide. During a test of filter flow rate, the influent valve to the filter is closed for 5 minutes. The water level drops 25 inches during this period. What is the filtration rate for the filter in gpm/sq ft?

ANS_____

7. A filter is 35 ft long and 25 ft wide. During a test of flow rate, the influent valve to the filter is closed for seven minutes. The water level drops 23 inches during this period. What is the filtration rate for the filter in gpm/sq ft?

ANS_____

PRACTICE PROBLEMS 11.3: Unit Filter Run Volume (UFRV)

1. The total water filtered between backwashes is 2.91 MG. If the filter is 20 ft by 20 ft, what is the unit filter run volume in gal/sq ft?

ANS_____

2. The total water filtered during a filter run is 4,140,000 gallons. If the filter is 30 ft long and 20 ft wide, what is the UFRV in gal/sq ft?

ANS_____

3. A total of 3,040,000 gallons of water are filtered during a particular filter run. If the filter is 25 ft long and 20 ft wide, what was the UFRV in gal/sq ft?

ANS_____

4. The average filtration rate for a filter was determined to be 3.3 gpm/sq ft. If the filter run time was 3260 minutes, what was the unit filter run volume in gal/sq ft?

ANS_____

5. A filter ran 62.6 hours between backwashes. If the average filtration rate during that time was 2.4 gpm/sq ft, what was the UFRV in gal/sq ft?

ANS_____

PRACTICE PROBLEMS 11.4: Backwash Rate, gpm/sq ft

1. A filter with a surface area of 375 sq ft has a backwash flow rate of 3440 gpm. What is the filter backwash rate in gpm/sq ft?

ANS_____

2. A filter 20 ft long and 15 ft wide has a backwash flow rate of 3760 gpm. What is the filter backwash rate in gpm/sq ft?

ANS_____

3. A filter has a backwash rate of 18 gpm/sq ft. What is this backwash rate expressed as in./min?

ANS_____

4. A filter 30 ft by 20 ft has a backwash flow rate of 3700 gpm. What is the filter backwash rate in gpm/sq ft?

ANS_____

5. A filter 20 ft long and 15 ft wide has a backwash rate of 3150 gpm. What is this backwash rate expressed as in./min rise?

ANS_____

PRACTICE PROBLEMS 11.5: Volume of Backwash Water Required, gal

1. A backwash flow rate of 6,700 gpm for a total backwashing period of 7 minutes would require how many gallons of water for backwashing?

ANS_____

2. For a backwash flow rate of 9,050 gpm and a total backwash time of 8 minutes, how many gallons of water will be required for backwashing?

ANS_____

3. How many gallons of water would be required to provide a backwash flow rate of 4860 gpm for a total of 6 minutes?

ANS_____

4. A backwash flow rate of 6,840 gpm for a total of 6 minutes would require how many gallons of water?

ANS_____

PRACTICE PROBLEMS 11.6: Req'd Depth of Backwash Water Tank, ft

1. A total of 58,940 gallons of water will be required to provide a 7-minute backwash of a filter. What depth of water is required in the backwash water tank to provide this backwashing capability? The tank has a diameter of 45 ft.

ANS_____

2. The volume of water required for backwashing has been calculated to be 61,700 gallons. What is the required depth of water in the backwash water tank to provide this amount of water if the diameter of the tank is 50 ft?

ANS_____

3. A total of 41,400 gallons of water will be required for backwashing a filter. What depth of water is required in the backwash water tank to provide this much water? The diameter of the tank is 40 ft.

ANS_____

4. A backwash rate of 7200 gpm is desired for a total backwash time of 8 minutes. What depth of water is required in the backwash water tank to provide this much water? The diameter of the tank is 50 ft.

ANS_____

5. A backwash rate of 8950 gpm is desired for a total backwash time of 7 minutes. What depth of water is required in the backwash water tank to provide this backwashing capability? The diameter of the tank is 45 ft.

ANS_____

PRACTICE PROBLEMS 11.7: Backwash Pumping Rate, gpm

1. A filter is 40 ft long and 20 ft wide. If the desired backwash rate is 18 gpm/sq ft, what backwash pumping rate (gpm) will be required?

ANS_____

2. The desired backwash pumping rate for a filter is 15 gpm/sq ft. If the filter is 35 ft long and 25 ft wide, what backwash pumping rate (gpm) will be required?

ANS_____

3. A filter is 20 ft square. If the desired backwash rate is 17 gpm/sq ft, what backwash pumping rate (gpm) will be required?

ANS_____

4. The desired backwash pumping rate for a filter is 24 gpm/sq ft. If the filter is 25 ft long and 20 ft wide, what backwash pumping rate (gpm) will be required?

ANS_____

PRACTICE PROBLEMS 11.8: Percent of Product Water Used
for Backwashing

1. A total of 16,840,000 gallons of water are filtered during a filter run. If 73,700 gallons of this product water are used for backwashing, what percent of the product water is used for backwashing?

ANS_____

2. A total of 5.86 MG of water were filtered during a filter run. If 36,300 gallons of this product water were used for backwashing, what percent of the product water was used for backwashing?

ANS_____

3. 58,200 gallons of product water are used for filter backwashing at the end of a filter run. If a total of 12,962,000 gallons are filtered during the filter run, what percent of the product water is used for backwashing?

ANS_____

4. A total of 10,905,000 gallons of water are filtered during a particular filter run. If 51,710 gallons of product water are used for backwashing, what percent of the product water is used for backwashing?

ANS_____

PRACTICE PROBLEMS 11.9: Percent Mud Ball Volume

1. A total 3475-mL sample of filter media was taken for mud ball evaluation. The volume of water in the graduated cylinder rose from 500 mL to 517 mL when mud balls were placed in the cylinder. What is the percent mud ball volume of the sample?

ANS_____

2. Five samples of filter media are taken for mud ball evaluation. The volume of water in the graduated cylinder rose from 500 mL to 529 mL when mud balls were placed in the cylinder. What is the percent mud ball volume of the sample? (The mud ball sampler has a volume of 695 mL.)

ANS_____

3. A filter media is tested for the presence of mud balls. The mud ball sampler has a total sample volume of 695 mL. Five samples were taken from the filter. When the mud balls were placed in 500 mL of water, the water level rose to 581 mL. What is the percent mud ball volume of the sample?

ANS_____

4. Five samples of media filter are taken and evaluated for the presence of mud balls. The volume of water in the graduated cylinder rises from 500 m*L* to 552 m*L* when the mud balls are placed in the water. What is the percent mud ball volume of the sample? (The mud ball sampler has a sample volume of 695 m*L*.)

ANS_____

CHAPTER 11—ACHIEVEMENT TEST

1. During a 75-hour filter run, a total of 12.9 million gallons of water are filtered. What is the average gpm flow rate through the filter during this time?

ANS_____

2. A filter is 35 ft long and 20 ft wide. If the filter receives a flow of 3.26 MGD, what is the filtration rate in gpm/sq ft?

ANS_____

3. The total water filtered between backwashes is 2.64 MG. If the filter is 20 ft by 20 ft, what is the unit filter run volume in gal/sq ft?

ANS_____

4. At an average flow rate through a filter of 2800 gpm, how long a filter run (in hours) would be required to produce 14.3 MG of filtered water?

ANS_____

5. A filter is 35 ft long and 25 ft wide. To verify the flow rate through the filter, the filter influent valve is closed for a period of 5 minutes and the water drop is measured. If the water level in the filter drops 15 inches during the 5-minute period, what is the gpm flow rate through the filter?

ANS_____

6. A total of 3,260,000 gallons of water are filtered during a particular filter run. If the filter is 25 ft long and 25 ft wide, what is the unit filter run volume in gal/sq ft?

ANS_____

7. A filter 35 ft long and 25 ft wide produces a total of 14,050,000 gallons during a 74.6-hr filter run. What is the average filtration rate (gpm/sq ft) for this filter run?

ANS_____

8. A filter with a surface area of 375 sq ft has a backwash flow rate of 3120 gpm. What is the filter backwash rate in gpm/sq ft?

ANS_____

CHAPTER 11—ACHIEVEMENT TEST—Cont'd

9. A backwash flow rate of 6,100 gpm for a total backwashing period of 6 minutes would require how many gallons of water for backwashing?

ANS_____

10. The influent valve to a filter is closed for a 5-minute period. During this time, the water level in the filter drops 14 inches. If the filter is 40 ft long and 25 ft wide, what is the filtration rate in gpm/sq ft?

ANS_____

11. A total of 51,600 gallons of water will be required to provide a 6-minute backwash of a filter. What depth of water is required in the backwash water tank to provide this backwashing capability? (The tank has a diameter of 50 ft.)

ANS_____

12. The average filtration rate for a filter was determined to be 3.1 gpm/sq ft. If the filter run time was 3510 minutes, what was the unit filter run volume in gal/sq ft?

ANS_____

13. A filter is 40 ft long and 20 ft wide. During a test of filter flow rate, the influent valve to the filter is closed for 5 minutes. The water level drops 22 inches during this period. What is the filtration rate for the filter in gpm/sq ft?

ANS_____

14. A filter 30 ft by 20 ft has a backwash flow rate of 3600 gpm. What is the filter backwash rate in gpm/sq ft?

ANS_____

15. How many gallons of water would be required to provide a backwash flow rate of 4700 gpm for a total of 8 minutes?

ANS_____

16. The desired backwash pumping rate for a filter is 15 gpm/sq ft. If the filter is 35 ft long and 25 ft wide, what backwash pumping rate (gpm) will be required?

ANS_____

CHAPTER 11—ACHIEVEMENT TEST—Cont'd

17. A filter 20 ft long and 20 ft wide has a backwash rate of 2900 gpm. What is this backwash rate expressed as in./min rise?

ANS_____

18. A filter is 35 ft long and 25 ft wide. During a test of flow rate, the influent valve to the filter is closed for seven minutes. The water level drops 19 inches during this period. What is the filtration rate for the filter in gpm/sq ft?

ANS_____

19. A filter is 40 ft long and 20 ft wide. If the desired backwash rate is 20 gpm/sq ft, what backwash pumping rate (gpm) will be required?

ANS_____

20. A total of 17,650,000 gallons of water are filtered during a filter run. If 69,200 gallons of this product water are used for backwashing, what percent of the product water is used for backwashing?

ANS_____

21. A total of 85,200 gallons of water will be required for backwashing a filter. What depth of water is required in the backwash water tank to provide this much water? The diameter of the tank is 40 ft.

ANS_____

22. A total 3475-mL sample of filter media was taken for mud ball evaluation. The volume of water in the graduated cylinder rose from 500 mL to 521 mL when the mud balls were placed in the cylinder. What is the percent mud ball volume of the sample?

ANS_____

23. 49,100 gallons of product water are used for filter backwashing at the end of a filter run. If a total of 13.7 MG are filtered during the filter run, what percent of the product water is used for backwashing?

ANS_____

24 Five samples of filter media are taken for mud ball evaluation. The volume of water in the water in the graduated cylinder rose from 500 mL to 568 mL when the mud balls were placed in the cylinder. What is the percent mud ball volume of the sample? (The mud ball sampler has a volume of 695 mL.)

ANS_____

12 Chlorination

PRACTICE PROBLEMS 12.1: Chlorine Feed Rate

1. Determine the chlorinator setting (lbs/day) needed to treat a flow of 3.2 MGD with a chlorine dose of 1.7 mg/L.

ANS_____

2. A flow of 1,260,000 gpd is to receive a chlorine dose of 2.4 mg/L. What should the chlorinator setting be, in lbs/day?

ANS_____

3. A pipeline 10 inches in diameter and 900 ft long is to be treated with a chlorine dose of 50 mg/L. How many lbs of chlorine will this require?

ANS_____

4. A chlorinator setting is 41 lbs per 24 hours. If the flow being treated is 3.02 MGD, what is the chlorine dosage expressed as mg/*L*?

ANS_____

5. The flow totalizer reading at 8 am on Wednesday was 18,762,102 gal and at 8 am on Thursday was 19,414,522 gal. If the chlorinator setting is 15 lbs for this 24-hour period, what is the chlorine dosage in mg/*L*?

ANS_____

PRACTICE PROBLEMS 12.2: Chlorine Dose, Demand and Residual

1. The chlorine demand of a water is 1.4 mg/L. If the desired chlorine residual is 0.5 mg/L, what is the desired chlorine dose, in mg/L?

ANS_____

2. The chlorine dosage for a water is 2.7 mg/L. If the chlorine residual after 30 minutes contact time is found to be 0.8 mg/L, what is the chlorine demand expressed in mg/L?

ANS_____

3. A flow of 3,790,000 gpd is to be disinfected with chlorine. If the chlorine demand is 2.4 mg/L and a chlorine residual of 0.7 mg/L is desired, what should be the chlorinator setting in lbs/day?

ANS_____

4. A chlorinator setting is increased by 5 lbs/day. The chlorine residual before the increased dosage was 0.4 mg/*L*. After the increased dose, the chlorine residual was 0.7 mg/*L*. The average flow rate being treated is 960,000 gpd. Is the water being chlorinated beyond the breakpoint?

ANS_____

5. A chlorinator setting of 18 lbs of chlorine per 24 hours results in a chlorine residual of 0.3 mg/*L*. The chlorinator setting is increased to 22 lbs per 24 hours. The chlorine residual increases to 0.5 mg/*L* at this new dosage rate. The average flow being treated is 1,980,000 gpd. On the basis of this data, is the water being chlorinated past the breakpoint?

ANS_____

PRACTICE PROBLEMS 12.3: Dry Hypochlorite Feed Rate

1. A chlorine dose of 46 lbs/day is required to treat a particular water. If calcium hypochlorite (65% available chlorine) is to be used, how many lbs/day of hypochlorite will be required?

ANS_____

2. A chlorine dose of 38 lbs/day is required to disinfect a flow of 2,145,000 gpd. If the calcium hypochlorite to be used contains 65% available chlorine, how many lbs/day hypochlorite will be required?

ANS_____

3. A water flow of 917,000 gpd requires a chlorine dose of 2.6 mg/*L*. If calcium hypochlorite (65% available chlorine) is to be used, how many lbs/day of hypochlorite are required?

ANS_____

4. A total of 51 lbs of hypochlorite (65% available chlorine) are used in a day. If the flow rate treated is 1,402,000 gpd, what is the chlorine dosage in mg/*L*?

ANS_____

5. A flow of 2,944,000 gpd is disinfected with a calcium hypochlorite (70% available chlorine). If 47 lbs of hypochlorite are used in a 24-hour period, what is the mg/*L* chlorine dosage?

ANS_____

PRACTICE PROBLEMS 12.4: Hypochlorite Solution Feed Rate

1. A total of 35 lbs/day sodium hypochlorite are required for disinfection of a 1.8 MGD flow. How many gallons per day sodium hypochlorite is this?

ANS_____

2. A chlorine dose of 2.9 mg/L is required for adequate disinfection of a water. If a flow of 795,000 gpd will be treated, how many gpd of sodium hypochlorite will be required? The sodium hypochlorite contains 13% available chlorine.

ANS_____

3. A hypochlorinator is used to disinfect the water pumped from a well. The hypochlorite solution contains 2% available chlorine. A chlorine dose of 2.1 mg/L is required for adequate disinfection throughout the distribution system. If the flow from the well is 220,000 gpd, how much sodium hypochlorite (gpd) will be required?

ANS_____

4. Water from a well is disinfected by a hypochlorinator. The flow totalizer indicates that 2,110,000 gallons of water were pumped during a 7-day period. The 2% sodium hypochlorite solution used to treat the well water is pumped from a 3-ft diameter storage tank. During the 7-day period, the level in the tank dropped 2 ft 9 in. What is the chlorine dosage in mg/L?

ANS_____

5. A hypochlorite solution (3% available chlorine) is used to disinfect a water. A chlorine dose of 1.6 mg/L is desired to maintain an adequate chlorine residual. If the flow being treated is 300 gpm, what hypochlorite solution flow (in gpd) will be required?

ANS_____

6. A sodium hypochlorite solution (2% available chlorine) is used to disinfect the water pumped from a well. A chlorine dose of 2.7 mg/L is required for adequate disinfection. How many gallons per day of sodium hypochlorite will be required if the flow being chlorinated is 940,000 gpd?

ANS_____

PRACTICE PROBLEMS 12.5: Percent Strength of Solution

1. A total of 20 lbs of calcium hypochlorite (65% available chlorine) are added to 55 gallons of water. What is the percent chlorine (by weight) of the solution?

ANS_____

2. If 300 grams of calcium hypochlorite are dissolved in 5 gallons of water, what is the percent chlorine (by weight) of the solution? (1 gram = 0.0022 lbs)

ANS_____

3. How many pounds of dry hypochlorite (70% available chlorine) must be added to 55 gallons of water to make a 2% chlorine solution?

ANS_____

4. A 12% liquid hypochlorite is to be used in making up a 2% hypochlorite solution. How many gallons of liquid hypochlorite should be mixed with water to produce 30 gallons of a 2% hypochlorite solution?

ANS_____

5. How many gallons of 14% liquid hypochlorite should be mixed with water to produce 100 gallons of a 1.3% hypochlorite solution?

ANS_____

6. If 5 gallons of a 13% sodium hypochlorite solution are added to 55-gallon drum, how much water should be added to the drum to produce a 3% hypochlorite solution?

ANS_____

PRACTICE PROBLEMS 12.6: Mixing Hypochlorite Solutions

1. If 40 lbs of a hypochlorite solution (12% available chlorine) are mixed with 250 lbs of another hypochlorite solution (1% available chlorine), what is the percent chlorine of the solution mixture?

ANS_____

2. If 8 gallons of a 10% hypochlorite solution are mixed with 50 gallons of a 1.5% hypochlorite solution, what is the percent strength of the solution mixture?

ANS_____

3. If 15 gallons of a 11% hypochlorite solution are added to 65 gallons of 1% hypochlorite solution, what is the percent strength of the solution mixture?

ANS_____

4. What weights of a 1.5% hypochlorite solution and a 13% hypochlorite solution must be mixed to make 800 lbs of a 3% solution?

ANS_____

5. What weights of a 1% solution and a 14% solution must be mixed to make 500 lbs of a 2% solution?

ANS_____

6. How many gallons of an 11% solution and water should be mixed together to form 400 gallons of a 2% solution?

ANS_____

PRACTICE PROBLEMS 12.7: Mixing Hypochlorite Solutions

1. The average calcium hypochlorite use at a plant is 41 lbs/day. If the chemical inventory in stock is 900 lbs, how many days' supply is this?

ANS_____

2. The average daily use of sodium hypochlorite solution at a plant is 70 gpd. A chemical feed tank has a diameter of 3 ft and contains solution to a depth of 3 ft 7 inches. How many days' supply is represented by the solution in the tank?

ANS_____

3. An average of 26 lbs of chorine are used each day at a plant. How many pounds of chlorine would be used in one week if the hour meter on the pump registered 140 hrs of operation that week?

ANS_____

4. A chlorine cylinder has 89 lbs of chlorine at the beginning of a week. The chlorinator setting is 14 lbs per 24 hours. If the pump hour meter indicates the pump has operated a total of 105 hrs during the week, how many lbs chlorine should be in the cylinder at the end of the week?

ANS_____

5. An average of 52 lbs of chlorine are used each day at a plant. How many 150-lb chlorine cylinders will be required each month? (Assume a 30-day month.)

ANS_____

6. The average sodium hypochlorite use at a plant is 60 gpd. If the chemical feed tank is 3 ft in diameter, how many feet should the solution level in the tank drop in two days' time?

ANS_____

CHAPTER 12—ACHIEVEMENT TEST

1. The chlorine demand of a water is 1.6 mg/*L*. If the desired chlorine residual is 0.8 mg/*L*, what is the desired chlorine dose, in mg/*L*?

ANS_____

2. Determine the chlorinator setting (lbs/day) needed to treat a flow of 910,000 gpd with a chlorine dose of 2.2 mg/*L*.

ANS_____

3. A chlorine dose of 50 lbs/day is required to treat a water. If calcium hypochlorite (65% available chorine) is to be used, how many lbs/day of hypochlorite will be required?

ANS_____

4. A total of 48 lbs/day sodium hypochlorite are required for disinfection of a 2.06 MGD flow. How many gallons per day sodium hypochlorite is this?

ANS_____

5. The chlorine dosage for a water is 2.9 mg/*L*. If the chlorine residual after 30 minutes contact time is found to be 0.6 mg/*L*, what is the chlorine demand expressed in mg/*L*?

ANS_____

6. A total of 25 lbs of calcium hypochlorite (65% available chlorine) are added to 62 gallons of water. What is the percent chlorine (by weight) of the solution?

ANS_____

7. What chlorinator setting is required to treat a flow of 1580 gpm with a chlorine dose of 2.6 mg/*L*?

ANS_____

8. A chlorine dose of 2.6 mg/*L* is required for adequate disinfection of a water. If a flow of 1.17 MGD will be treated, how many gpd of sodium hypochlorite will be required? The sodium hypochlorite contains 12% available chlorine.

ANS_____

CHAPTER 12—ACHIEVEMENT TEST—Cont'd

9. A pipeline 6 inches in diameter and 1500 ft long is to be treated with a chlorine dose of 50 mg/L. How many lbs of chlorine will this require?

ANS_____

10. A chlorinator setting of 14 lbs of chlorine per 24 hours results in a chlorine residual of 0.5 mg/L. The chlorinator setting is increased to 18 lbs per 24 hours. The chlorine residual increases to 0.7 mg/L at this new dosage rate. The average flow being treated is 2,070,000 gpd. On the basis of this data, is the water being chlorinated past the breakpoint?

ANS_____

11. If 60 gal of a 12% hypochlorite solution are mixed with 240 gal of a 2% hypochlorite solution, what is the percent strength of the solution mixture?

ANS_____

12. The average calcium hypochlorite use at a plant is 36 lbs/day. If the chemical inventory in stock is 280 lbs, how many days' supply is this?

ANS_____

13. The flow totalizer reading a 8 am on Thursday was 42,197,660 gal and at 8 am on Friday was 43,981,669 gal. If the chlorinator setting is 19 lbs for this 24-hr period, what is the chorine dosage in mg/*L*?

ANS_____

14. A chlorine dose of 35 lbs/day is required to disinfect a flow of 1,980,000 gpd. If the calcium hypochlorite to be used contains 70% available chlorine, how many lbs/day hypochlorite will be required?

ANS_____

15. Water from a well is disinfected by a hypochlorinator. The flow totalizer indicates that 2,417,000 gallons of water were pumped during a 7-day period. The 2% sodium hypochlorite solution used to treat the well water is pumped from a 3-ft diameter storage tank. During the 7-day period, the level in the tank dropped 3 ft 2 in. What is the chlorine dosage in mg/*L*?

ANS_____

16. A flow of 3,150,000 gpd is to be disinfected with chlorine. If the chlorine demand is 2.3 mg/*L* and a chlorine residual of 0.5 mg/*L* is desired, what should be the chlorinator setting in lbs/day?

ANS_____

CHAPTER 12—ACHIEVEMENT TEST—Cont'd

17. If 10 gallons of a 14% hypochlorite solution are mixed with 40 gallons of a 1% hypochlorite solution, what is the percent strength of the solution mixture?

ANS_____

18. A total of 65 lbs of hypochlorite (65% available chlorine) are used in a day. If the flow rate treated is 1,750,000 gpd, what is the chlorine dosage in mg/L?

ANS_____

19. A hypochlorite solution (3% available chlorine) is used to disinfect a water. A chlorine dose of 2.4 mg/L is desired to maintain an adequate chlorine residual. If the flow being treated is 350 gpm, what hypochlorite solution flow (in gpd) will be required?

ANS_____

20. The average daily use of sodium hypochlorite at a plant is 85 gpd. The chemical feed tank has a diameter of 3 ft and contains solution to a depth of 3 ft 11 in. How many days' supply are represented by the solution in the tank?

ANS_____

21. How many pounds of dry hypochlorite (65% available chlorine) must be added to 75 gallons of water to make a 2% chlorine solution?

ANS_____

22. An average of 28 lbs of chlorine are used each day at a plant. How many pound of chlorine would be used in one week if the hour meter on the pump registers 135 hrs of operation that week?

ANS_____

23. An average of 45 lbs of chlorine are used each day at a plant. How many 150-lb chlorine cylinders will be required each month? (Assume a 30-day month.)

ANS_____

24. How many gallons of 12% liquid hypochlorite should be mixed with water to produce 200 gallons of a 1.5% hypochlorite solution?

ANS_____

13 *Fluoridation*

PRACTICE PROBLEMS 13.1: Methods of Expressing Concentration

1. Express 2.4% concentration in terms of mg/*L* concentration.

ANS_____

2. Convert 6700 mg/*L* to percent.

ANS_____

3. Express 22% concentration in terms of mg/*L*.

ANS_____

4. Express 14 lbs/MG concentration as mg/*L*.

ANS_____

5. Convert 1.8 mg/*L* to lbs/MG.

ANS_____

6. Express 23 lbs/MG concentration as mg/*L* concentration.

ANS_____

PRACTICE PROBLEMS 13.2: Percent Fluoride Ion in a Compound

1. Calculate the percent fluoride ion present in hydrofluosilicic acid, H_2SiF_6.
   (The atomic weights are as follows: H = 1.008; Si = 28.06; F = 19.00)

   ANS_____

2. Calculate the percent fluoride ion present in sodium fluoride, NaF.
   (The atomic weights are as follows: Na = 22.997; F = 19.00)

   ANS_____

3. Calculate the percent fluoride ion present in sodium silicofluoride, Na_2SiF_6.
   The atomic weights are as follows: Na = 22.997; Si = 28.06; F = 19.00)

   ANS_____

4. Calculate the percent fluoride ion present in magnesium silicofluoride, $MgSiF_6$.
(Atomic weights are as follows: Magnesium = 24.32; Si = 28.06; F = 19.00)

ANS_____

PRACTICE PROBLEMS 13.3: Calculating Dry Feed Rate, lbs/day

1. A fluoride dosage of 1.5 mg/L is desired. The flow to be treated is 980,000 gpd. How many lbs/day dry sodium silicofluoride (Na_2SiF_6) will be required if the commercial purity of the Na_2SiF_6 is 98% and the percent of fluoride ion in the compound is 60.6%? (Assume the raw water contains no fluoride.)

ANS_____

2. A fluoride dosage of 1.2 mg/L is desired. How many lbs/day dry sodium silicofluoride (Na_2SiF_6) will be required if the flow to be treated is 1.68 MGD? The commercial purity of the sodium silicofluoride is 98% and the percent of fluoride ion in Na_2SiF_6 is 60.6%. (Assume the water to be treated contains no fluoride.)

ANS_____

3. A flow of 2,796,000 gpd is to be treated with sodium silicofluoride, Na_2SiF_6. The raw water contains no fluoride. If the desired fluoride concentration in the water is 1.6 mg/L, what should be the chemical feed rate (in lbs/day)? The manufacturer's data indicates that each pound of Na_2SiF_6 contains 0.6 lbs of fluoride ion. (Assume the raw water contains no fluoride.)

ANS_____

4. A flow of 3.06 MGD is to be treated with sodium fluoride, NaF. The raw water contains no fluoride and the desired fluoride concentration in the finished water is 1.0 mg/*L*. What should be the chemical feed rate in lbs/day? (Manufacturer's data indicates that each pound of NaF contains 0.44 lbs of fluoride ion.)

ANS_____

5. A flow of 775,000 gpd is to be treated with sodium fluoride, NaF. The raw water contains 0.09 mg/*L* fluoride and the desired fluoride level in the finished water is 1.3 mg/*L*. What should be the chemical feed rate in lbs/day? (Manufacturer's data indicates that each pound of NaF contains 0.45 lbs of fluoride ion.)

ANS_____

PRACTICE PROBLEMS 13.4: Percent Strength of a Solution

1. If 8 lbs of sodium fluoride (NaF) are mixed with 50 gal of water what is the percent strength of the solution? (The commercially available sodium fluoride is 98% pure.)

ANS_____

2. If 15 lbs of sodium fluoride, NaF, are dissolved in 75 gallons of water, what is the percent strength of the solution? (Assume the chemical used is 100% pure NaF.)

ANS_____

3. How many pound of sodium fluoride (98% pure) must be added to 200 gallons of water to make a 1.5% solution of sodium fluoride?

ANS_____

4. If 10 lbs of sodium fluoride are mixed with 50 gal water, what is the percent strength of the solution? (The commercial sodium fluoride is 98% pure.)

ANS_____

5. How many pounds of sodium fluoride (98% pure) must be added to 150 gallons of water to make a 2% solution of sodium fluoride?

ANS_____

PRACTICE PROBLEMS 13.5: Calculating Solution Feed Rates

1. A flow of 4.12 MGD is to be treated with a 22% solution of hydrofluosilicic acid. The acid has a specific gravity of 1.2. If the desired fluoride level in the finished water is 1.1 mg/*L*, what should be the solution feed rate (in gpd)? The raw water contains no fluoride. (The percent fluoride ion content of H_2SiF_6 is 79.2%.)

ANS_____

2. A flow of 2.8 MGD is to be treated with a 20% solution of hydrofluosilicic acid (H_2SiF_6). The desired fluoride concentration is 1.1 mg/L. What should be the solution feed rate (in gpd)? The hydrofluosilicic acid weighs 9.8 lbs/gal. (Assume the water to be treated contains no natural fluoride level.) The percent fluoride ion content of H_2SiF_6 is 79.2%.

ANS_____

3. A flow of 840,000 gpd is to be treated with a 2.1% solution of sodium fluoride, NaF. If the desired fluoride ion concentration is 1.6 mg/*L*, what should be the sodium fluoride feed rate (in gpd)? Sodium fluoride has a fluoride ion content of 45.25%. The water to be treated contains 0.08 mg/*L* fluoride ion. (Assume the solution density is 8.34 lbs/day.)

ANS_____

4. A flow of 1,475,000 gpd is to be treated with a 2.2% solution of sodium fluoride, NaF. The desired fluoride level in the finished water is 1.5 mg/*L*. What should be the sodium fluoride solution feed rate (in gpd)? Sodium fluoride has a fluoride ion content of 45.25%. The raw water contains no fluoride. (Assume the solution density is 8.34 lbs/gal.)

ANS_____

5. The desired solution feed rate has been determined to be 90 gpd. What is this feed rate expressed as m*L*/min?

ANS_____

6. A flow of 2.68 MGD is to be treated with a 25% solution of hydrofluosilicic acid, H_2SiF_6. The raw water contains no fluoride and the desired fluoride concentration is 1.0 mg/*L*. The hydrofluosilicic acid weighs 9.8 lbs/gal. What should be the m*L*/min solution feed rate? (The percent fluoride content of H_2SiF_6 is 79.2%.)

ANS_____

PRACTICE PROBLEMS 13.6: Feed Rates Using Charts and Nomographs

1. A flow of 7 MGD is to be treated with a 25% solution of hydrofluosilicic acid, H_2SiF_6. The fluoride ion concentration to be added is 1.3 mg/L. Use the Treatment Chart I given in Appendix 3 to determine the appropriate solution feed rate.

ANS_____

2. A flow of 20 MGD is to be treated with a 20% hydrofluosilicic acid, H_2SiF_6. The desired fluoride ion concentration in the finished water is 1.5 mg/L and the raw water contains no fluoride. Use the Treatment Chart II given in Appendix 3 to determine the appropriate solution feed rate.

ANS_____

3. A flow of 700,000 gpd is to be treated with a saturated solution of sodium fluoride, NaF. Use the fluoridation nomograph in Appendix 4 to determine the gpd sodium fluoride required to provide a fluoride level of 0.9 mg/L. (Assume the raw water contains no fluoride.)

ANS_____

4. A flow of 4 MGD is to be treated with a 20% solution of hydrofluosilicic acid, H_2SiF_6. The desired fluoride concentration is 1.1 mg/*L*. The raw water contains a fluoride level of 0.07 mg/*L*. Use the Treatment Chart I in Appendix 3 to determine the appropriate solution feed rate.

ANS_____

5. A flow of 1.5 MGD is to be treated with a 3% solution of sodium fluoride. The raw water has a natural fluoride level of 0.1 mg/*L* and the desired fluoride level in the finished water is 1.6 mg/*L*. Use the Treatment Chart IV in Appendix 3 to determine the sodium fluoride feed rate, in gpd.

ANS_____

6. A flow of 400 gpm is to be treated with a 30% solution of hydrofluosilicic acid, H_2SiF_6. If the desired fluoride concentration of the finished water is 1.1 mg/*L* and the raw water contains no fluoride, what is the required solution feed rate in gpd? Use the nomograph provided in Appendix 4.

ANS_____

PRACTICE PROBLEMS 13.7: Calculating mg/*L* Fluoride Dosage

1. A total of 30 lbs/day of sodium silicofluoride (Na_2SiF_6) were added to a flow of 1,480,000 gpd. The commercial purity of the sodium silicofluoride was 98% and the percent fluoride ion content of Na_2SiF_6 is 60.7%. What was the concentration of fluoride ion in the treated water?

ANS_____

2. A flow of 320,000 gpd is treated with 5 lbs/day sodium fluoride, NaF. The commercial purity of the sodium fluoride is 98% and the fluoride ion content of NaF is 45.25%. Under these conditions what is the fluoride ion dosage in mg/*L*?

ANS_____

3. A flow of 3.76 MGD is treated with a 20% solution of hydrofluosilicic acid, H_2SiF_6. If the solution feed rate is 25 gpd, what is the calculated fluoride ion concentration of the treated water? Assume the acid weighs 9.8 lbs/gal and the percent fluoride ion in H_2SiF_6 is 79.2%. The raw water contains no fluoride.

ANS_____

4. A flow of 1,840,000 gpd is treated with a 10% solution of hydrofluosilicic acid, H_2SiF_6. If the solution feed rate is 25 gpd, what is the calculated fluoride ion concentration of the finished water? The acid weighs 9.04 lbs/gal and the percent fluoride ion in H_2SiF_6 is 79.2%. The raw water contains no fluoride.

ANS_____

5. A flow of 2,620,000 gpd is to be treated with a 4% saturated solution of sodium fluoride, NaF. If the solution feed rate is 120 gpd, what is the calculated fluoride ion level in the finished water? Assume the solution weighs 8.34 lbs/gal. The percent fluoride ion in NaF is 45.25%. The raw water contains 0.1 mg/L fluoride.

ANS_____

PRACTICE PROBLEMS 13.8: Solution Mixtures

1. A tank contains 500 lbs of 15% hydrofluosilicic acid, H_2SiF_6. If 2400 lbs of 25% hydrofluosilicic acid are added to the tank, what is the percent strength of the solution mixture?

ANS_____

2. If 800 lbs of 25% hydrofluosilicic acid were added to a tank containing 250 lbs of 15% hydrofluosilicic acid, what would be the percent strength of the solution mixture?

ANS_____

3. A tank of 18% hydrofluosilicic acid (H_2SiF_6) contains 350 gallons. If a truck delivers 2000 gallons of 20% hydrofluosilicic acid to be added to the tank, what is the percent strength of the solution mixture? Assume the 20% solution weighs 9.8 lbs/gal and the 18% solution weighs 9.6 lbs/gal.

ANS_____

4. A tank contains 250 gallons of a 10% hydrofluosilicic acid. If 1000 gallons of a 20% hydrofluosilicic acid are added to the tank, what is the percent strength of the solution mixture? Assume the 10% acid weighs 9.04 lbs/gal and the 20% acid weighs 9.8 lbs/gal.

ANS_____

5. A tank contains 200 gallons of a 9% hydrofluosilicic acid with a specific gravity of 1.075. If a total of 1500 gallons of 15% hydrofluosilicic acid are added to the tank, what is the percent strength of the solution mixture? Assume the 15% acid solution weighs 9.4 lbs/gal.

ANS_____

CHAPTER 13—ACHIEVEMENT TEST

1. Express 2.8% concentration in terms of mg/*L* concentration.

ANS_____

2. Calculate the percent fluoride ion present in hydrofluosilicic acid, H_2SiF_6.
(The atomic weights are as follows: H = 1.008; Si = 28.06; F = 19.00)

ANS_____

3. Express 24 lbs/MG concentration as mg/*L* concentration.

ANS_____

4. A fluoride ion dosage of 1.4 mg/*L* is desired. The flow to be treated is 1,156,000
gpd. How many lbs/day dry sodium silicofluoride (Na_2SiF_6) will be required if the
commercial purity of the Na_2SiF_6 is 98% and the percent of fluoride ion in the
compound is 60.6%? (Assume the raw water contains no fluoride.)

ANS_____

5. Calculate the percent fluoride ion present in sodium fluoride, NaF.
(The atomic weights are as follows: Na = 22.997; F = 19.00)

ANS_____

6. If 70 lbs of sodium fluoride (NaF) are mixed with 500 gal of water, what is the
percent strength of the solution? (The commercially available sodium fluoride is
98% pure.)

ANS_____

7. Convert 25,000 mg/*L* to percent.

ANS_____

8. The desired solution feed rate has been determined to be 75 gpd. What is this
feed rate expressed as m*L*/min?

ANS_____

CHAPTER 13—ACHIEVEMENT TEST—Cont'd

9. A fluoride dosage of 1.3 mg/*L* is desired. How many lbs/day dry sodium fluoride (NaF) will be required if the flow to be treated is 2.17 MGD? The commercial purity of the sodium fluoride is 98% and the percent of fluoride ion in NaF is 45.25%. (Assume the water to be treated contains no fluoride.)

ANS_____

10. How many pounds of sodium fluoride (98% pure) must be added to 400 gallons of water to make a 2% solution of sodium fluoride?

ANS_____

11. A flow of 3.96 MGD is to be treated with a 22% solution of hydrofluosilicic acid. The acid has a specific gravity of 1.2. If the desired fluoride level in the finished water is 1.5 mg/*L*, what should be the solution feed rate (in gpd)? The raw water contains no fluoride. (The percent fluoride ion content of H_2SiF_6 is 79.2%.)

ANS_____

12. If 25 lbs of sodium fluoride are mixed with 120 gallons water, what is the percent strength of the solution? (The commercial sodium fluoride is 98% pure.)

ANS_____

13. A flow of 7 MGD is to be treated with a 20% solution of hydrofluosilicic acid, H_2SiF_6. The fluoride ion concentration to be added is 1.4 mg/L. Use the Treatment Chart I given in Appendix 3 to determine the appropriate solution feed rate.

ANS_____

14. A flow of 1,772,000 gpd is to be treated with sodium fluoride, NaF. The raw water contains 0.08 mg/L fluoride and the desired fluoride level in the finished water is 1.3 mg/L. What should be the chemical feed rate in lbs/day? (Manufacturer's data indicates that each pound of NaF contains 0.44 lbs of fluoride ion.)

ANS_____

CHAPTER 13—ACHIEVEMENT TEST—Cont'd

15. A flow of 2.6 MGD is to be treated with a 20% solution of hydrofluosilicic acid (H_2SiF_6). The desired fluoride concentration is 1.2 mg/L. What should be the solution feed rate (in gpd)? The hydrofluosilicic acid weighs 9.8 lbs/gal. (Assume the water to be treated contains no natural fluoride level.) The percent fluoride ion content of H_2SiF_6 is 79.2%.

ANS_____

16. A tank contains 400 lbs of 15% hydrofluosilicic acid, H_2SiF_6. If 1500 lbs of 20% hydrofluosilicic acid are added to the tank, what is the percent strength of the solution mixture?

ANS_____

17. A total of 38 lbs/day of sodium silicofluoride (Na_2SiF_6) were added to a flow of 1,070,000 gpd. The commercial purity of the sodium silicofluoride was 98% and the percent fluoride ion content of Na_2SiF_6 is 60.7%. What was the concentration of fluoride ion in the treated water?

ANS_____

18. A flow of 1300 gpm is treated with a 10% solution of hydrofluosilicic acid, H_2SiF_6. If the solution feed rate is 30 gpd, what is the calculated fluoride ion concentration of the finished water? The acid weighs 9.04 lbs/gal and the percent fluoride ion in H_2SiF_6 is 79.2%. The raw water contains no fluoride.

ANS_____

19. A flow of 300 gpm is to be treated with a saturated solution of sodium fluoride, NaF. Use the fluoridation nomograph in Appendix 4 to determine the gpd sodium fluoride required to provide a fluoride level of 1.1 mg/*L*. (Assume the raw water contains no fluoride.)

ANS_____

20. A tank contains 215 gallons of a 10% hydrofluosilicic acid. If 500 gallons of a 20% hydrofluosilicic acid are added to the tank, what is the percent strength of the solution mixture? Assume the 10% acid weighs 9.04 lbs/gal and the 20% acid weighs 9.8 lbs/gal.

ANS_____

CHAPTER 13—ACHIEVEMENT TEST—Cont'd

21. A flow of 2.08 MGD is to be treated with a 25% solution of hydrofluosilicic acid, H_2SiF_6. The raw water contains no fluoride and the desired fluoride concentration is 1.1 mg/L. The hydrofluosilicic acid weighs 9.8 lbs/gal. What should be the mL/min solution feed rate? (The percent fluoride content of H_2SiF_6 is 79.2%.)

ANS_____

22. A tank contains 128 gallons of a 9% hydrofluosilicic acid with a specific gravity of 1.075. If a total of 800 gallons of 15% hydrofluosilicic acid are added to the tank, what is the percent strength of the solution mixture? Assume the 15% acid solution weighs 9.4 lbs/gal.

ANS_____

23. A flow of 2 MGD is to be treated with a 3% solution of sodium fluoride. The raw water has a natural fluoride level of 0.1 mg/L and the desired fluoride level in the finished water is 1.4 mg/L. Use the Treatment Chart IV in Appendix 3 to determine the sodium fluoride feed rate, in gpd.

ANS_____

24. A flow of 2,880,000 gpd is to be treated with a 4% saturated solution of sodium fluoride, NaF. If the solution feed rate is 125 gpd, what is the calculated fluoride ion level in the finished water? Assume the solution weighs 8.34 lbs/gal. The percent fluoride ion in NaF is 45.25%. The raw water contains 0.1 mg/*L* fluoride.

ANS_____

14 *Softening*

PRACTICE PROBLEMS 14.1: Equivalent Weight

1. The atomic weight of sodium is 22.997. If sodium has a valence of 1, what is the equivalent weight of sodium?

ANS_____

2. Calcium has an atomic weight of 40.08 and a valence of 2. What is the equivalent weight of calcium?

ANS_____

3. The atomic weight of magnesium is 24.32. If the valence of magnesium is 2, what is the equivalent weight of magnesium?

ANS_____

4. Given the atomic weights listed below, what is the equivalent weight of sodium chloride (NaCl)? (Na = 22.997; Cl = 35.457)

ANS_____

5. Given the atomic weights listed below, what is the equivalent weight of calcium carbonate ($CaCO_3$)? (Ca = 40.08; C = 12.010; O = 16.000)

ANS_____

6. 275 milligrams of calcium are equal to how many milliequivalents of calcium? (The equivalent weight of calcium is 20.04.)

ANS_____

7. The magnesium content of a water is 32 mg/L. How many milliequivalents/liter of magnesium is this?

ANS_____

8. Given a calcium content of 30 mg/L, how many milliequivalents/liter of calcium is this?

ANS_____

PRACTICE PROBLEMS 14.2: Hardness as CaCO$_3$

1. The calcium content of a water sample is 37 mg/L. What is this calcium hardness expressed as CaCO3? (The equivalent weight of calcium is 20.04 and the equivalent weight of CaCO3 is 50.045.)

ANS_____

2. The magnesium content of a water is 31 mg/L. What is this magnesium hardness expressed as CaCO3? (The equivalent weight of magnesium is 12.16 and the equivalent weight of CaCO3 is 50.045.)

ANS_____

3. A water contains 19 mg/L calcium. What is this calcium hardness expressed as CaCO3? (The equivalent weight of calcium is 20.04 and CaCO3 is 50.045.)

ANS_____

4. A water has a calcium concentration of 72 mg/*L* as $CaCO_3$ and a magnesium concentration of 89 mg/*L* as $CaCO_3$. What is the total hardness (as $CaCO_3$) of the sample?

ANS_____

5. Determine the total hardness as $CaCO_3$ of a water that has a calcium content of 25 mg/*L* and a magnesium content of 9 mg/*L*. (The equivalent weight of calcium is 20.04, of magnesium is 12.16, and of $CaCO_3$ is 50.045.)

ANS_____

6. Determine the total hardness as $CaCO_3$ of a water that has a calcium content of 19 mg/*L* and a magnesium content of 14 mg/*L*. (The equivalent weight of calcium is 20.04, of magnesium is 12.16, and of $CaCO_3$ is 50.045.)

ANS_____

PRACTICE PROBLEMS 14.3: Carbonate and Noncarbonate Hardness

1. A sample of water contains 120 mg/L alkalinity as $CaCO_3$ and 105 mg/L total hardness as $CaCO_3$. What is the carbonate and noncarbonate hardness of this water?

ANS_____

2. The alkalinity of a water is 95 mg/L as $CaCO_3$. If the total hardness of the water is 118 mg/L as $CaCO_3$, what is the carbonate and noncarbonate hardness in mg/L as $CaCO_3$?

ANS_____

3. A water has an alkalinity of 82 mg/L as $CaCO_3$ and a total hardness of 114 mg/L. What is the carbonate and noncarbonate hardness of the water?

ANS_____

4. A water sample contains 110 mg/*L* alkalinity as $CaCO_3$ and 97 mg/*L* total hardness as $CaCO_3$. What is the carbonate and noncarbonate hardness of this water?

ANS_____

5. The alkalinity of a water is 101 mg/*L* as $CaCO_3$. If the total hardness of the water is 119 mg/*L* as $CaCO_3$, what is the carbonate and noncarbonate hardness in mg/*L* as $CaCO_3$?

ANS_____

PRACTICE PROBLEMS 14.4: Phenolphthalein and Total Alkalinity

1. A 100-mL water sample is tested for phenolphthalein alkalinity. If 1 mL titrant is used to pH 8.3 and the sulfuric acid solution has a normality of 0.02N, what is the phenolphthalein alkalinity of the water? (mg/L as $CaCO_3$)

ANS_____

2. A 100-mL water sample is tested for phenolphthalein alkalinity. If 1.3 mL titrant is used to pH 8.3 and the normality of the sulfuric acid solution is 0.02N, what is the phenolphthalein alkalinity of the water? (mg/L as $CaCO_3$)

ANS_____

3. A 100-mL sample of water is tested for alkalinity. The normality of the sulfuric acid used for titrating is 0.02N. If 0.2 mL titrant is used to pH 8.3, and 6.9 mL titrant is used to pH 4.5, what is the phenolphthalein and total alkalinity of the sample?

ANS_____

4. A 100-m*L* sample of water is tested for phenolphthalein and total alkalinity. A total of 0 m*L* titrant is used to pH 8.3 and a total of 7.1 m*L* titrant is used to titrate to pH 4.5. The normality of the acid used for titrating is 0.02 *N*. What is the phenolphthalein and total alkalinity of the sample? (mg/*L* as CaCO₃)

ANS_____

5. A 100-m*L* sample of water is tested for alkalinity. The normality of the sulfuric acid used for titrating is 0.02*N*. If 0.4 m*L* titrant is used to pH 8.3, and 5.9 m*L* titrant is used to pH 4.5, what is the phenolphthalein and total alkalinity of the sample?

ANS_____

PRACTICE PROBLEMS 14.5: Bicarbonate, Carbonate and Hydroxide Alkalinity

Alkalinity, mg/L as CaCO3			
Results of Titration	Bicarbonate Alkalinity	Carbonate Alkalinity	Hydroxide Alkalinity
P = O	T	O	O
P is less than 1/2 T	T – 2P	2P	O
P = 1/2 T	O	2P	O
P is greater then 1/2 T	O	2T – 2P	2P – T
P = T	O	O	T

Where P = Phenolphthalein alkalinity
       T = Total alkalinity

1. A water sample is tested for phenolphthalein and total alkalinity. If the phenolphthalein alkalinity is 9 mg/L as CaCO3 and the total alkalinity is 48 mg/L as CaCO3, what is the bicarbonate, carbonate, and hydroxide alkalinity of the water?

ANS_____

2. A water sample is found to have a phenolphthalein alkalinity of 0 mg/L and a total alkalinity of 68 mg/L. What is the bicarbonate, carbonate and hydroxide alkalinity of the water?

ANS_____

3. The phenolphthalein alkalinity of a water sample is 14 mg/L as $CaCO_3$ and the total alkalinity is 22 mg/L as $CaCO_3$. What is the bicarbonate, carbonate and hydroxide alkalinity of the water?

ANS_____

4. Alkalinity titrations on a 100-mL water sample resulted as follows: 1.2 mL titrant used to pH 8.3; 5.1 mL total titrant used to pH 4.5. The normality of the sulfuric acid was 0.02N. What is the phenolphthalein, total, bicarbonate, carbonate and hydroxide alkalinity of the water?

ANS_____

5. Alkalinity titrations on a 100-mL water sample resulted as follows: 1.4 mL titrant used to pH 8.3; 2.8 mL total titrant used to pH 4.5. The normality of the sulfuric acid was 0.02N. What is the phenolphthalein, total, bicarbonate, carbonate and hydroxide alkalinity of the water?

ANS_____

PRACTICE PROBLEMS 14.6: Lime Dosage for Softening

1. A water sample has a carbon dioxide content of 7 mg/L as CO_2, total alkalinity of 140 mg/L as $CaCO_3$, and magnesium content of 25 mg/L as MG^{+2}. Approximately how much quicklime (CaO) (90% purity) will be required for softening? (Assume 15% excess lime.)

ANS_____

2. The characteristics of a water are as follows: 4 mg/L CO_2 as CO_2, 162 mg/L total alkalinity as $CaCO_3$, and 15 mg/L magnesium as Mg^{+2}. What is the estimated hydrated lime (Ca(OH)$_2$) (90% pure) dosage in mg/L required for softening? (Assume 15% excess lime.)

ANS_____

3. A water sample has the following characteristics: 5 mg/L CO_2 as CO_2, 108 mg/L total alkalinity as $CaCO_3$, and 11 mg/L magnesium as Mg^{+2}. What is the estimated hydrated lime (Ca(OH)$_2$) (90% purity) dosage in mg/L required for softening? (Assume 15% excess lime.)

ANS_____

4. A water sample has a carbon dioxide content of 8 mg/L as CO_2 , total alkalinity of 170 mg/L as $CaCO_3$, and magnesium content of 17 mg/L as MG^{+2}. Approximately how much quicklime (CaO) (90% purity) will be required for softening?

ANS_____

PRACTICE PROBLEMS 14.7: Soda Ash Dosage

1. A water has a total hardness of 240 mg/L as $CaCO_3$ and a total alkalinity of 173 mg/L. What soda ash dosage (mg/L) will be required to remove the noncarbonate hardness?

ANS_____

2. The alkalinity of a water is 109 mg/L as $CaCO_3$ and the total hardness is 225 mg/L as $CaCO_3$. What soda ash dosage (in mg/L) is required to remove the noncarbonate hardness?

ANS_____

3. A water sample has a total hardness of 258 mg/L as $CaCO_3$ and a total alkalinity of 160 mg/L. What soda ash dosage (mg/L) will be required to remove the noncarbonate hardness?

ANS_____

4. Calculate the soda ash required (in mg/*L*) to soften a water if the water has a total hardness of 218 mg/*L* and a total alkalinity of 110 mg/*L*.

ANS_____

PRACTICE PROBLEMS 14.8: Carbon Dioxide for Recarbonation

1. The A, B, C and D factors of the excess lime equation have been calculated as follows:
   A = 10 mg/L; B = 132 mg/L; C = O; D = 65 mg/L. If the residual magnesium is 5 mg/L,
   what is the carbon dioxide dosage (in mg/L) required for recarbonation?

   ANS_____

2. The A, B, C and D factors of the excess lime equation have been calculated as:
   A = 9 mg/L; B = 92 mg/L; C = 6; D = 110 mg/L. If the residual magnesium is
   4 mg/L, what carbon dioxide dosage would be required for recarbonation?

   ANS_____

3. The A, B, C and D factors of the excess lime equation were determined to be as follows:
   A = 8 mg/L; B = 111 mg/L; C = 2; D = 54 mg/L. The magnesium residual is 6 mg/L.
   What is the carbon dioxide dosage required for recarbonation?

   ANS_____

4. The A, B, C and D factors of the excess lime equation have determined to be: A = 7 mg/*L*; B = 115 mg/*L*; C = 7; D = 47 mg/*L*. If the residual magnesium is 3 mg/*L*, what carbon dioxide dosage would be required for recarbonation?

ANS_____

PRACTICE PROBLEMS 14.9: Chemical Feeder Settings

1. Jar tests indicate that the optimum lime dosage is 190 mg/*L*. If the flow to be treated is 2.46 MGD, what should be the chemical feeder setting in lbs/day?

ANS_____

2. The optimum lime dosage for a water has been determined to be 185 mg/*L*. If the flow to be treated is 3,140,000 gpd, what should be the chemical feeder setting in lbs/day and lbs/min?

ANS_____

3. A soda ash dosage of 65 mg/*L* is required to remove noncarbonate hardness. What should be the lbs/hr chemical feeder setting if the flow rate to be treated is 4.55 MGD?

ANS_____

4. What should the chemical feeder setting be, in lbs/day and lbs/min, if the optimum lime dosage has been determined to be 135 mg/L and the flow to be treated is 1,970,000 gpd?

ANS_____

5. A total of 40 mg/L soda ash is required to remove noncarbonate hardness from a water. What should be the chemical feeder setting in lbs/hr and lbs/min if the flow to be treated is 3,050,000 gpd?

ANS_____

PRACTICE PROBLEMS 14.10: Expressions of Hardness, mg/*L* and gpg

1. The total hardness of a water is 216 mg/*L*. What is this hardness expressed as grains per gallon?

   ANS_____

2. The total hardness of a water is 12.8 mg/*L*. What is this concentration expressed as mg/*L*?

   ANS_____

3. The total hardness of a water is reported as 245 mg/*L*. What is this hardness expressed as grains per gallon?

   ANS_____

4. A hardness of 16 gpg is equivalent to how many mg/L?

ANS_____

PRACTICE PROBLEMS 14.11: Ion Exchange Capacity

1. The hardness removal capacity of an ion exchange resin is 24,000 grains/cu ft. If the softener contains a total of 110 cu ft of resin, what is the exchange capacity of the softener (in grains) ?

ANS_____

2. An ion exchange water softener has a diameter of 5 ft. The depth of resin is 3.5 ft. If the resin has a removal capacity of 20 kilograins/cu ft, what is the exchange capacity of the softener (in grains)?

ANS_____

3. The hardness removal capacity of an exchange resin is 22 kilograins/cu ft. If the softener contains a total of 280 cu ft of resin, what is the exchange capacity of the softener (in grains)?

ANS_____

4. An ion exchange water softener has a diameter of 6 ft. The depth of resin is 4 ft. If the resin has a removal capacity of 20 kilograins/cu ft, what is the exchange capacity of the softener (in grains)?

ANS_____

PRACTICE PROBLEMS 14.12: Water Treatment Capacity

1. An ion exchange softener has an exchange capacity of 2,192,000 grains. If the hardness of the water to be treated is 17.8 gpg, how many gallons of water can be treated before regeneration of the resin is required?

ANS_____

2. The exchange capacity of an ion exchange softener is 4,500,000 grains. If the hardness of the water to be treated is 16.2 grains/gallon, how many gallons of water can be treated before regeneration of the resin is required?

ANS_____

3. An ion exchange softener has an exchange capacity of 3,877,000 grains. If the hardness of the water is 265 mg/L, how many gallons of water can be treated before regeneration of the resin is required?

ANS_____

4. The hardness removal capacity of an ion exchange resin is 20 kilograins/cu ft. The softener contains a total of 180 cu ft of resin. If the water to be treated contains 14.9 gpg hardness, how many gallons of water can be treated before regeneration of the resin is required?

ANS_____

5. The hardness removal capacity of an ion exchange resin is 21,000 grains/cu ft. The softener has a diameter of 4 ft and a depth of resin of 2.5 ft. If the water to be treated contains 13.7 gpg hardness, how many gallons of water can be treated before regeneration of the resin is required?

ANS_____

PRACTICE PROBLEMS 14.13: Operating Time

1. An ion exchange softener can treat a total of 590,000 gallons of water before regeneration is required. If the flow rate treated is 24,500 gph how many hours of operation are there before regeneration is required?

ANS_____

2. An ion exchange softener can treat a total of 782,000 gallons of water before regeneration of the resin is required. If the water is to be treated at a rate of 25,000 gph, how many hours of operation are there until regeneration is required?

ANS_____

3. A total of 356,000 gallons of water can be treated by an ion exchange water softener before regeneration of the resin is required. If the flow rate to be treated is 225 gpm, what is the operating time (in hrs) until regeneration of the resin will be required?

ANS_____

4. The exchange capacity of an ion exchange softener is 3,158,000 grains. The water to be treated contains 13 gpg total hardness. If the flow rate to be treated is 210 gpm, how many hours of operation are there until regeneration of the resin will be required?

ANS_____

5. The exchange capacity of an ion exchange softener is 3,780,000 grains. The water to be treated contains 11.4 gpg total hardness. If the flow rate to be treated is 289,000 gpd, how many hours of operation are there until regeneration of the resin will be required?

ANS_____

PRACTICE PROBLEMS 14.14: Salt and Brine Required

Salt Solutions		
% NaCl	lbs NaCl / gal	lbs NaCl / cu ft
10	0.874	6.69
11	0.990	7.41
12	1.09	8.14
13	1.19	8.83
14	1.29	9.63
15	1.39	10.4

1. An ion exchange softener will remove 2,300,000 grains hardness from the water until the resin must be regenerated. If 0.4 lbs salt are required for each kilograin removed, how many pounds of salt will be required for preparing the brine to be used in resin regeneration?

ANS_____

2. A total of 1,330,000 grains hardness are removed by an ion exchange softener before the resin must be regenerated. If 0.3 lbs salt is required for each kilograin removed, how many pounds of salt will be required for preparing the brine to be used in resin regeneration?

ANS_____

3. 395 lbs salt are required in making up a brine solution for regeneration. If the brine solution is to be a 13% solution of salt, how many gallons of brine will be required for regeneration of the softener? (Use the Salt Solution table to determine the lbs salt/gal brine for a 13% brine solution.)

ANS_____

4. A total of 410 lbs salt will be required to regenerate an ion exchange softener. If the brine solution is to be a 14% brine solution, how many gallons brine will be required? (Use the Salt Solutions table to determine the lbs salt/gal brine for a 14% brine solution.)

ANS_____

5. An ion exchange softener removes 1,310,000 grains hardness from the water before the resin must be regenerated and 0.4 lbs salt is required for each kilograin hardness removed. If the brine solution is to be a 12% brine solution, how many gallons of brine will be required for regeneration of the softener? (Use the Salt Solutions table to determine the lbs salt/gal brine for a 12% brine solution.)

ANS_____

CHAPTER 14—Achievement Test

1. Sodium has an atomic weight of 22.997 and a valence of 1. What is the equivalent weight of sodium?

ANS_____

2. The calcium content of a water sample is 42 mg/L. What is this calcium hardness expressed as $CaCO_3$? (The equivalent weight of calcium is 20.04 and the equivalent weight of $CaCO_3$ is 50.045.)

ANS_____

3. A 100-mL sample of water is tested for phenolphthalein alkalinity. If 1.6 mL titrant is used to pH 8.3 and the normality of the sulfuric acid solution is 0.02 N, what is the phenolphthalein alkalinity of the water? (mg/L as $CaCO_3$)

ANS_____

4. A water sample is tested for alkalinity. If the phenolphthalein alkalinity is 10 mg/L as $CaCO_3$ and the total alkalinity is 52 mg/L as $CaCO_3$, what is the bicarbonate, carbonate, and hydroxide alkalinity of the water? (Use the alkalinity table in Appendix 5.)

ANS_____

5. Given the atomic weights listed below, what is the equivalent weight of magnesium chloride ($MgCl_2$)? ($Mg = 24.32$; $Cl = 35.457$) The valence of magnesium is 2.

ANS_____

6. The magnesium content of a water sample is 28 mg/L. What is this magnesium hardness expressed as $CaCO_3$? (The equivalent weight of magnesium is 12.16 and the equivalent weight of $CaCO_3$ is 50.045.)

ANS_____

7. 22 milligrams of magnesium are equal to how many milliequivalents of calcium?

ANS_____

8. The characteristics of a water are as follows: 7 mg/L CO_2 as CO_2 , 116 mg/L total alkalinity as $CaCO_3$, 14 mg/L magnesium as Mg^{+2}. What is the estimated hydrated lime ($Ca(OH)_2$) (90% pure) dosage in mg/L required for softening?

ANS_____

CHAPTER 14—Achievement Test—Cont'd

9. Determine the total hardness as $CaCO_3$ of a water that has a calcium content of 30 mg/L and a magnesium content of 12 mg/L. (The equivalent weight of calcium is 20.04, of magnesium is 12.16, and of $CaCO_3$ is 50.045.)

ANS_____

10. A sample of water contains 115 mg/L alkalinity as $CaCO_3$ and 103 mg/L total hardness as $CaCO_3$. What is the carbonate and noncarbonate hardness of this water?

ANS_____

11. A water sample has a carbon dioxide content of 4 mg/L as CO_2 , total alkalinity of 155 mg/L as $CaCO_3$, and magnesium content of 12 mg/L as MG^{+2}. Approximately how much quicklime (CaO) (90% purity) will be required for softening? (Assume 15% excess lime.)

ANS_____

12. A 100-m*L* water sample is tested for phenolphthalein and total alkalinity. A total of 0.4 m*L* titrant is used to pH 8.3 and a total of 6.7 m*L* titrant is used to titrate to pH 4.5. The normality of sulfuric acid is 0.02*N*. What is the phenolphthalein and total alkalinity of the sample? (mg/*L* as $CaCO_3$)

ANS_____

13. A water has an alkalinity of 85 mg/*L* as $CaCO_3$ and a total hardness of 118 mg/*L*. What is the carbonate and noncarbonate hardness of the water?

ANS_____

14. The alkalinity of a water is 95 mg/*L* as $CaCO_3$ and the total hardness is 214 mg/*L* as $CaCO_3$. What soda ash dosage (in mg/*L*) is required to remove the noncarbonate hardness?

ANS_____

CHAPTER 14—Achievement Test—Cont'd

15. The phenolphthalein alkalinity of a water sample is 15 mg/L as $CaCO_3$ and the total alkalinity is 24 mg/L as $CaCO_3$. What is the bicarbonate, carbonate and hydroxide alkalinity of the water? (Use the alkalinity table provided in Appendix 5.)

ANS_____

16. The hardness removal capacity of an exchange resin is 21,000 grains/cu ft. If the softener contains a total of 90 cu ft of resin, what is the exchange capacity of the softener (in grains)?

ANS_____

17. A water sample has a total hardness of 255 mg/L as $CaCO_3$ and a total alkalinity of 158 mg/L. What soda ash dosage (mg/L) will be required to remove the noncarbonate hardness?

ANS_____

18. A hardness of 14 gpg is equivalent to how many mg/*L*?

ANS_____

19. The A, B, C and D factors of the excess lime equation have been calculated as follows: A. = 8 mg/*L*; B = 140 mg/*L*; C = 3; D = 63 mg/*L*. If the residual magnesium is 7 mg/*L*, what is the carbon dioxide dosage (in mg/*L*) required for recarbonation? (Assume a 15% excess lime.)

ANS_____

20. An ion exchange softener has an exchange capacity of 2,207,000 grains. If the hardness of the water to be treated is 17.6 gpg, how many gallons of water can be treated before regeneration of the resin is required?

ANS_____

CHAPTER 14—Achievement Test—Cont'd

21. An ion exchange water softener has a diameter of 6 ft. The depth of resin is 4 ft. If the resin has a removal capacity of 19 kilograins/cu ft, what is the total exchange capacity of the softener (in grains)?

ANS_____

22. The A, B, C and D factors of the excess lime equation have been calculated as: A = 8 mg/L; B = 90 mg/L; C = 7; D = 105 mg/L. If the residual magnesium is 5 mg/L, what carbon dioxide dosage would be required for recarbonation? (Assume a 15% excess lime.)

ANS_____

23. The hardness removal capacity of an ion exchange resin is 21 kilograins/cu ft. The softener contains a total of 190 cu ft of resin. If the water to be treated contains 15.1 gpg hardness, how many gallons of water can be treated before regeneration of the resin is required?

ANS_____

24. The optimum lime dosage for a water has been determined to be 170 mg/*L*. If the flow to be treated is 2,065,000 gpd, what should be the chemical feeder setting in lbs/day and lbs/min?

ANS_____

25. The total hardness of a water is 229 mg/*L*. What is this hardness expressed as grains per gallon?

ANS_____

26. An ion exchange softener can treat a total of 583,000 gallons of water before regeneration of the resin is required. If the flow rate treated is 24,720 gph, how many hours of operation are there until regeneration is required?

ANS_____

CHAPTER 14—Achievement Test—Cont'd

27. An ion exchange softener will remove 2,268,000 grains hardness from the water until the resin must be regenerated. If 0.3 lbs salt is required for each kilograin hardness removed, how many pounds of salt will be required for preparing the brine to be used in resin regeneration?

ANS_____

28. A soda ash dosage of 60 mg/L is required to remove noncarbonate hardness. What should be the chemical feeder setting if the flow rate to be treated is 3.15 MGD?

ANS_____

29. 409 lbs salt are required in making up the brine solution for regeneration of the resin. If the brine solution is to be an 11% solution of salt, how many gallons of brine will be required for regeneration of the softener? (Use the Salt Solutions table given on the next page.)

ANS_____

30. A total of 352,000 gallons of water can be treated by an ion exchange water softener before regeneration of the resin is required. If the flow rate to be treated is 230 gpm, what is the operating time (in hrs) until regeneration of the resin will be required?

ANS_____

Salt Solutions		
% NaCl	lbs NaCl / gal	lbs NaCl / cu ft
10	0.874	6.69
11	0.990	7.41
12	1.09	8.14
13	1.19	8.83
14	1.29	9.63
15	1.39	10.4

15 *Laboratory*

PRACTICE PROBLEMS 15.1: Estimating Flow From A Faucet

1. The flow from a faucet filled up a one-gallon container in 72 seconds. What was the gpm flow rate from the faucet?

ANS_____

2. A one-gallon container was filled in 56 seconds. What was the gpm flow rate from the faucet?

ANS_____

3. The flow from a faucet filled a one-gallon container in 1 minute 38 seconds. What was the gpm flow rate from the faucet?

ANS_____

4. At a flow rate of 0.6 gpm, how many seconds will it take to fill the one gallon container?

ANS_____

5. How many seconds will it take to fill up a one-gallon container if the flow rate from the faucet is 0.4 gpm?

ANS_____

PRACTICE PROBLEMS 15.2: Service Line Flushing Time

1. How many minutes will it take to flush a 30-ft length of 3/4-inch diameter service line, if the flow rate through the line is 0.5 gpm?

ANS_____

2. How many minutes will it take to flush a 25-ft length of 3/4-inch diameter service line, if the flow rate through the line is 0.6 gpm?

ANS_____

3. At a flow rate of 0.5 gpm, how long (minutes and seconds) will it take to flush a 35-ft length of 3/4-inch service line?

ANS_____

4. How long (minutes and seconds) will it take to flush a 20-ft length of 1/2-inch line if the flow rate through the line is 0.5 gpm?

ANS_____

PRACTICE PROBLEMS 15.3: Solution Concentration

1. If 2.3 equivalents of a chemical are dissolved is 1.5 liters solution, what is the normality of the solution?

ANS_____

2. A 700-m*L* solution contains 1.2 equivalents of a chemical. What is the normality of the solution?

ANS_____

3. How many milliliters of 0.5*N* NaOH will react with 400 m*L* of 0.02*N* HCl?

ANS_____

4. To prepare a sodium arsenite solution, 5 grams $NaAsO_2$ are to be dissolved in distilled water and diluted to 1 liter. If 4.72 grams $NaAsO_2$ are used in preparing the solution, how many milliliters solution should be prepared?

ANS_____

5. To prepare a methyl red indicator solution, 100 mg methyl red sodium salt are to be dissolved and diluted to 100 mL. If 97.24 mg of methyl red sodium salt are used to prepare the solution, how many milliliters solution should be prepared?

ANS_____

6. To prepare a copper sulfate solution, 100 grams $CuSO_4 • 5 H_2O$ are to be dissolved in distilled water and diluted to 1 liter. If 93.42 grams $CuSO_4 • 5 H_2O$ are used in preparing the solution, how many milliliters solution should be prepared?

ANS_____

PRACTICE PROBLEMS 15.4: Chlorine Residual

1. The chlorine residual indicated by the color change is 0.4 mg/*L*. If 3 drops of sample water were used until the color was produced in the 10-m*L* distilled water, what is the estimated actual chlorine residual of the sample?

ANS_____

2. The chlorine residual indicated by the color change produced during the drop-dilution method is 0.3 mg/*L*. If 2 drops of sample water were used until the color was produced in the 10 m*L* distilled water, what is the estimated actual chlorine residual of the sample?

ANS_____

3. The drop-dilution method is used to estimate the actual chlorine residual of a sample. The chlorine residual indicated by the test is 0.2 mg/*L*. If 3 drops of sample water were required to produce the color change in the 10 m*L* distilled water, what is the estimated actual chlorine residual of the sample?

ANS_____

4. Four drops of sample water are required to produce a color change in the 10 mL distilled water during the drop-dilution method. The chlorine residual indicated by the test is 0.4 mg/*L*. What is the estimated actual chlorine residual of the water tested?

ANS_____

PRACTICE PROBLEMS 15.5: Temperature

1. The temperature of a water is 68°F. What is the water temperature expressed as °C?

ANS_____

2. Convert a water temperature of 14°C to °F.

ANS_____

3. The water temperature is 56°F. What is the temperature expressed in °C?

ANS_____

4. A water temperature of 11°C is equal to how many°F?

ANS_____

CHAPTER 15 ACHIEVEMENT TEST

1. How many minutes will it take to flush a 50-ft length of 3/4-inch diameter service line if the flow through the line is 0.75 gpm?

ANS_____

2. The flow from a faucet filled up the gallon container in 40 seconds. What was the gpm flow rate from the faucet?

ANS_____

3. If 2.2 equivalents of a chemical are dissolved in 1.3 liters solution, what is the normality of the solution?

ANS_____

4. The flow from a faucet filled up the gallon container in 1 minute 8 seconds. What was the gpm flow rate from the faucet?

ANS_____

5. The chlorine residual indicated by the color change is 0.3 mg/L. If 2 drops of sample water were used until the color was produced in the 10 mL distilled water, what is the estimated actual chlorine residual of the sample?

ANS_____

6. How long (minutes) will it take to flush a 70-ft length of 3/4-inch diameter service line if the flow through the line is 0.5 gpm?

ANS_____

CHAPTER 15 ACHIEVEMENT TEST—Cont'd

7. To prepare a particular standard solution of sodium hydroxide (NaOH) 50 grams NaOH are to be dissolved in water and diluted to 1 liter. If 44.19 grams are weighed out, how many milliliters solution should be prepared?

ANS_____

8. At a flow rate of 0.7 gpm, how long (minutes) should it take to fill the one-gallon container?

ANS_____

9. A 750-m*L* solution contains 1.1 equivalents of a chemical. What is the normality of the solution?

ANS_____

10. The chlorine residual indicated by the color change produced during the drop-dilution method is 0.4 mg/*L*. If 3 drops of sample water were used until the color was produced in the 10 m*L* distilled water, what is the estimated actual chlorine residual of the sample?

ANS_____

11. The influent to a treatment plant has a temperature of 72° F. What is this temperature expressed in degrees Celsius?

ANS_____

12. At a flow rate of 0.5 gpm, how long (minutes and seconds) will it take to flush a 50-ft length of 3/4-inch service line?

ANS_____

CHAPTER 15 ACHIEVEMENT TEST—Cont'd

13. *Standard Methods* indicates that a 0.0192N solution of silver nitrate ($AgNO_3$) is to be prepared using 3.27 grams $AgNO_3$ dissolved in 1 liter distilled water. If 2.96 grams of $AgNO_3$ are used to prepare the solution, how many milliliters distilled water should be used in making the solution?

ANS_____

14. How many milliliters of 0.2*N* NaOH will react with 400 ml of 0.01*N* HCl?

ANS_____

15. Four drops of sample water are required to produce a color change in the 10 m*L* distilled water during the drop-dilution method of estimating chlorine residual. The chlorine residual indicated by the test is 0.4 mg/*L*. What is the estimated actual chlorine residual of the water tested?

ANS_____

16. To prepare a standard solution, the directions indicate that 7.6992 grams of chemical are to be weighted out and diluted to one liter. If 7.2725 grams of chemical are used in making the solution, how many milliliters solution should be prepared?

ANS_____

17. The effluent of a treatment plant is 18° C. What is this expressed in degrees Fahrenheit?

ANS_____

18. The influent to a treatment plant has a temperature of 80° F. What is this temperature expressed in degrees Celsius?

ANS_____

Answer Key

Water Workbook
Answer Key

Chapter 1

PRACTICE PROBLEMS 1.1

1. (0.785)(80 ft)(80 ft)(30 ft)(7.48 gal/cu ft) = 1,127,386 gal

2. (70 ft)(12 ft)(20 ft) = 16,800 cu ft

3. (25 ft)(80 ft)(13 ft)(7.48 gal/cu ft) = 194,480 gal

4. (15 ft)(30 ft)(10 ft)(7.48 gal/cu ft) = 33,660 gal

5. (0.785)(50 ft)(50 ft)(14 ft)(7.48 gal/cu ft) = 205,513 gal

PRACTICE PROBLEMS 1.2

1. (3 ft)(5 ft)(350 ft) = 5250 cu ft

2. (0.785)(0.83 ft)(0.83 ft)(1500 ft)(7.48 gal/cu ft) = 6068 gal

3. $\dfrac{(5\ ft + 10\ ft)}{2}$(3 ft)(700 ft)(7.48 gal/cu ft) = 117,810 gal

4. (0.785)(0.5 ft)(0.5 ft)(1778 ft)(7.48 gal/cu ft) = 2610 gal

5. (5 ft)(3.7 ft)(1000 ft)(7.48 gal/cu ft) = 138,380 gal

PRACTICE PROBLEMS 1.3

1. $\dfrac{(3\ ft)(3.5\ ft)(600\ ft)}{27\ cu\ ft/cu\ yd}$ = 233 cu yds

2. (430 ft)(660 ft)(6 ft) = 1,702,800 cu ft

3. (200 yds)(1 yd)(1.33 yds) = 266 cu yds

4. $\dfrac{(5\ ft + 9\ ft)}{2}$(4.5 ft)(720 ft) = 22,680 cu ft

5. (750 ft)(2 ft)(2 ft) = 3000 cu ft

CHAPTER 1 ACHIEVEMENT TEST

1. (600 ft)(3.5 ft)(6 ft) = 12,600 cu ft

2. (0.785)(70 ft)(70 ft)(23 ft)(7.48 gal/cu ft) = 661,752 gal

3. (430 ft)(660 ft)(4 ft) = 1,135,200 cu ft

4. (80 ft)(15 ft)(20 ft) = 24,000 cu ft

5. (0.785)(0.67 ft)(0.67 ft)(3500 ft)(7.48 gal/cu ft) = 9225 gal

6. (900 ft)(2 ft)(2.5 ft) = 4500 cu ft

7. $\dfrac{(2.5\ ft)(3\ ft)(1500\ ft)}{27\ cu\ ft/cu\ yd}$ = 417 cu yds

8. (5.5 ft)(2.5 ft)(1000 ft)(7.48 gal/cu ft) = 102,850 gal

9. (20 ft)(60 ft)(13 ft)(7.48 gal/cu ft) = 116,688 gal

10. (6 ft)(3.7 ft)(2000 ft)(7.48 gal/cu ft) = 332,112 gal

11. (7 ft)(3.5 ft)(851 ft) = 20,850 gal

12. (0.785)(50 ft)(50 ft)(16 ft)(7.48 gal/cu ft) = 234,872 gal

Chapter 2

PRACTICE PROBLEMS 2.1

1. (2.6 ft)(3.5 ft)(2.2 fps)(60 sec/min) = 1201 cfm

2. (15 ft)(10 ft)(0.7 fpm)(7.48 gal/cu ft) = 785 gpm

3. $\dfrac{(3.5\ ft + 5.5\ ft)}{2}$(3.2 ft)(125 fpm) = 1800 cfm

4. (0.785)(0.5 ft)(0.5 ft)(2.6 fps)(7.48 gal/cu ft)(60 sec/min) = 229 gpm

5. (0.785)(2 ft)(2 ft)(4.3 fpm)(7.48 gal/cu ft) = 101 gpm

6. (0.785)(0.83 ft)(0.83 ft)(3.2 fps)(7.48 gal/cu ft)(60 sec/min)(0.5) = 388 gpm

PRACTICE PROBLEMS 2.2

1. (5 ft)(2.3 ft)(x fps)(60 sec/min)(7.48 gal/cu ft) = 13,400 gpm
   x = 2.6 ft

2. (0.785)(0.67 ft)(0.67 ft)(x fps)(7.48 gal/cu ft)(60 sec/min) = 537 gpm
   x = 3.4 fps

3. 500 ft/208 sec = 2.4 fps

4. (0.785)(0.83 ft)(0.83 ft)(2.6 fps) = (0.785)(0.67 ft)(0.67 ft)(x fps)
   x = 4 fps

5. 400 ft/88 sec = 4.5 fps

6. (0.785)(0.67 ft)(0.67 ft)(3.6 fps) = (0.785)(0.83 ft)(0.83 ft)(x fps)
   x = 2.3 fps

PRACTICE PROBLEMS 2.3

1. 35.3 MGD/7 = 5 MGD

2. 117.3 MG/30 days = 3.9 MGD

3. 428.2 MG/91 days = 4.7 MGD

4. 3,140,000 gal/1440 min = 2180 gpm

PRACTICE PROBLEMS 2.4

1. (5 cfs)(7.48 gal/cu ft)(60 sec/min) = 2244 gpm

2. (38 gps)(60 sec/min)(1440 min/day) = 3,283,200 gpd

3. $\dfrac{4,270,000 \text{ gpd}}{(1440 \text{ min/day})(7.48 \text{ gal/cu ft})}$ = 396 cfm

4. (5.6 MGD)(1.55 cfs/MGD) = 8.7 cfs

5. $\dfrac{(423,690 \text{ cfd})(7.48 \text{ gal/cu ft})}{1440 \text{ min/day}}$ = 2201 gpm

6. (2730 gpm)(1440 min/day) = 3,931,200 gpd

CHAPTER 2 ACHIEVEMENT TEST

1. (6 ft)(2.6 ft)(x fps)(7.48 gal/cu ft)(60 sec/min) = 15,500 gpm
   x = 2.2 fps

2. 27.8 MGD/7 = 4 MGD

3. (4.2 ft)(3.2 ft)(3.9 fps)(60 sec/min) = 3145 cfm

4. 377.6 MG/92 days = 4.1 MGD

5. (10 ft)(10 ft)(0.67 fpm)(7.48 gal/cu ft) = 501 gpm

6. (0.785)(0.67 ft)(0.67 ft)(x fps)(7.48 gal/cu ft)(60 sec/min) = 490 gpm
   x = 3 fps

7. (8 cfs)(7.48 gal/cu ft)(60 sec/min) = 3590 gpm

8. 127.6 MG/31 days = 4.1 MGD

9. (4.8 MGD)(1.55 cfs/MGD) = 7.4 cfs

10. (0.785)(2 ft)(2 ft)(3 fpm)(7.48 gal/cu ft) = 70 gpm

11. $\dfrac{(1,780,000 \text{ gpd})}{(1440 \text{ min/day})(7.48 \text{ gal/cu ft})}$ = 165 cfm

12. (0.785)(0.5 ft)(0.5 ft)(2.7 fps)(7.48 gal/cu ft)(60 sec/min) = 238 gpm

13. 300 ft/86 sec = 3.5 fps

14. (0.785)(0.83 ft)(0.83 ft)(2.4 fps) = (0.785)(0.67 ft)(0.67 ft)(x fps)
    x = 3.7 fps

15. (2150 gpm)(1440 min/day) = 3,096,000 gpd

16. 4,620,000 gal/1440 min = 3208 gpm

Chapter 3

PRACTICE PROBLEMS 3.1

1. (2.3 mg/L)(5.1 MGD)(8.34 lbs/gal) = 98 lbs/day

2. (5.9 mg/L)(3.8 MGD)(8.34 lbs/gal) = 187 lbs/day

3. $\dfrac{(8 \text{ mg}/L)(1.6 \text{ MGD})(8.34 \text{ lbs/gal})}{0.65}$ = 164 lbs/day

PRACTICE PROBLEMS 3.1—Cont'd

4. (10.5 mg/*L*)(4.6 MGD)(8.34 lbs/gal) = 403 lbs/day

5. (4.1 mg/*L*)(6.14 MGD)(8.34 lbs/gal) = 210 lbs/day

6. (50 mg/*L*)(0.085 MGD)(8.34 lbs/gal) = 35 lbs

7. (2120 mg/*L*)(0.224 MG)(8.34 lbs/gal) = 3960 lbs

8. $\dfrac{(8\ mg/L)(0.72\ MGD)(8.34\ lbs/gal)}{0.65}$ = 74 lbs/day

PRACTICE PROBLEMS 3.2

1. (425 mg/*L*)(1.62 MGD)(8.34 lbs/gal) = 5742 lbs/day

2. (26 mg/*L*)(2.98 MGD)(8.34 lbs/gal) = 646 lbs/day

3. (280 mg/*L*)(5.34 MGD)(8.34 lbs/gal) = 12,470 lbs/day

4. (135 mg/*L*)(3.54 MGD)(8.34 lbs/gal) = 3986 lbs/day

5. (295 mg/*L*)(3.30 MGD)(8.34 lbs/gal) = 8119 lbs/day

PRACTICE PROBLEMS 3.3

1. (148 mg/*L*)(5.2 MGD)(8.34 lbs/gal) = 6418 lbs/day SS

2. (189 mg/*L*)(1.89 MGD)(8.34 lbs/gal) = 2979 lbs/day SS

3. (136 mg/*L*)(4.79 MGD)(8.34 lbs/gal) = 5433 lbs/day BOD

4. (221 mg/*L*)(2.37 MGD)(8.34 lbs/gal) = 4368 solids

5. (118 mg/*L*)(4.14 MGD)(8.34 lbs/gal) = 4074 lbs/day

PRACTICE PROBLEMS 3.4

1. (2140 mg/*L*)(0.35 MG)(8.34 lbs/gal) = 6247 lbs SS

2. (1960 mg/*L*)(0.37 MG)(8.34 lbs/gal) = 6048 lbs MLVSS

3. (2960 mg/*L*)(0.18 MG)(8.34 lbs/gal) = 4444 lbs MLVSS

4. (2440 mg/*L*)(0.61 MG)(8.34 lbs/gal) = 12,413 lbs MLSS

5. (2890 mg/*L*)(0.49 MG)(8.34 lbs/gal) = 11,810 lbs MLSS

PRACTICE PROBLEMS 3.5

1. $(6210 \text{ mg}/L)(x \text{ MGD})(8.34 \text{ lbs/gal}) = 5300 \text{ lbs/day}$
   $x = 0.10 \text{ MGD}$

2. a) $(5970 \text{ mg}/L)(x \text{ MGD})(8.34 \text{ lbs/gal}) = 4600 \text{ lbs/day}$
   $x = 0.09 \text{ MGD}$

   b) 90,000 gpd/1440 min/day = 62.5 gpm

3. $(6540 \text{ mg}/L)(x \text{ MGD})(8.34 \text{ lbs/gal}) = 6090 \text{ lbs/day}$
   $x = 0.11 \text{ MGD}$

   Then 110,000 gpd ÷ 1440 min/day = 73 gpm

4. $(6280 \text{ mg}/L)(x \text{ MGD})(8.34 \text{ lbs/gal}) = 7400 \text{ lbs/day}$
   $x = 0.14 \text{ MGD}$

   Then 140,000 gpd ÷ 1440 min/day = 97 gpm

5. $(7140 \text{ mg}/L)(x \text{ MGD})(8.34 \text{ lbs/gal}) = 5700 \text{ lbs/day}$
   $x = 0.10 \text{ MGD}$

   Then 100,000 gpd ÷ 1440 min/day = 69 gpm

CHAPTER 3 ACHIEVEMENT TEST

1. $(2.4 \text{ mg}/L)(3.82 \text{ MGD})(8.34 \text{ lbs/gal}) = 76 \text{ lbs/day}$

2. $(15 \text{ mg}/L)(2.05 \text{ MGD})(8.34 \text{ lbs/gal}) = 256 \text{ lbs/day BOD}$

3. $(185 \text{ mg}/L)(4.6 \text{ MGD})(8.34 \text{ lbs/gal}) = 7097 \text{ lbs/day SS Rem.}$

4. $(9.9 \text{ mg}/L)(5.6 \text{ MGD})(8.34 \text{ lbs/gal}) = 462 \text{ lbs/day}$

5. $(315 \text{ mg}/L)(3.7 \text{ MGD})(8.34 \text{ lbs/gal}) = 9720 \text{ lbs/day SS}$

6. $\dfrac{(12 \text{ mg}/L)(2.8 \text{ MGD})(8.34 \text{ lbs/gal})}{0.65} = 431 \text{ lbs/day Hypochlorite}$

7. $(190 \text{ mg}/L)(3.22 \text{ MGD})(8.34 \text{ lbs/gal}) = 5102 \text{ lbs/day Solids}$

8. $(50 \text{ mg}/L)(0.08 \text{ MG})(8.34 \text{ lbs/gal}) = 33 \text{ lbs Chlorine}$

9. $(2610 \text{ mg}/L)(0.39 \text{ MG})(8.34 \text{ lbs/gal}) = 8489 \text{ lbs MLSS}$

10. $(5980 \text{ mg}/L)(x \text{ MGD})(8.34 \text{ lbs/gal}) = 5540 \text{ lbs/day}$
    $x = 0.11 \text{ MGD}$

CHAPTER 3 ACHIEVEMENT TEST—Cont'd

11. (125 mg/*L*)(3.168 MGD)(8.34 lbs/gal) = 3303 lbs/day COD

12. (235 mg/*L*)(3.15 MGD)(8.34 lbs/gal) = 6174 lbs BOD

13. (205 mg/*L*)(1.9 MGD)(8.34 lbs/gal) = 3248 lbs/day BOD Removed

14. (*x* mg/*L*)(5.1 MGD)(8.34 lbs/gal) = 340 lbs/day
    x = 8 mg/*L*

15. (5800 mg/*L*)(*x* MGD)(8.34 lbs/gal) = 6240 lbs/day
    x = 0.13 MGD
    Then, 130,000 gpd ÷ 1440 min/day = 90 gpm

Chapter 4

PRACTICE PROBLEMS 4.1

1. $\dfrac{3,000,000 \text{ gpd}}{(0.785)(90 \text{ ft})(90 \text{ ft})}$ = 472 gpd/sq ft

2. $\dfrac{4,320,000 \text{ gpd}}{(0.785)(80 \text{ ft})(80 \text{ ft})}$ = 860 gpd/sq ft

3. $\dfrac{3,600,000 \text{ gpd}}{850,000 \text{ sq ft}}$ = 4.2 gpd/sq ft

4. $\dfrac{264,706 \text{ cu ft/day}}{653,400 \text{ sq ft}}$ = 0.4 ft/day

   Then (0.4 ft/day)(12 in./ft) = 4.8 in./day

5. $\dfrac{5,080,000 \text{ gpd}}{(0.785)(85 \text{ ft})(85 \text{ ft})}$ = 896 gpd/sq ft

6. $\dfrac{4.3 \text{ ac-ft/day}}{20 \text{ ac}}$ = 0.22 ft/day or 3 in./day

PRACTICE PROBLEMS 4.2

1. $\dfrac{2,140,000 \text{ gpd}}{(80 \text{ ft})(30 \text{ ft})}$ = 892 gpd/sq ft

2. $\dfrac{2,620,000 \text{ gpd}}{(0.785)(70 \text{ ft})(70 \text{ ft})} = 681 \text{ gpd/sq ft}$

3. $\dfrac{3,280,000 \text{ gpd}}{(100 \text{ ft})(40 \text{ ft})} = 820 \text{ gpd/sq ft}$

4. $\dfrac{1,480,000 \text{ gpd}}{(20 \text{ ft})(80 \text{ ft})} = 925 \text{ gpd/sq ft}$

5. $\dfrac{2,360,000 \text{ gpd}}{(0.785)(60 \text{ ft})(60 \text{ ft})} = 835 \text{ gpd/sq ft}$

PRACTICE PROBLEMS 4.3

1. $\dfrac{2150 \text{ gpm}}{(30 \text{ ft})(25 \text{ ft})} = 2.9 \text{ gpm/sq ft}$

2. $\dfrac{3080 \text{ gpm}}{(40 \text{ ft})(20 \text{ ft})} = 3.85 \text{ gpm/sq ft}$

3. $\dfrac{2000 \text{ gpm}}{(25 \text{ ft})(50 \text{ ft})} = 1.6 \text{ gpm/sq ft}$

4. $\dfrac{1528 \text{ gpm}}{(45 \text{ ft})(25 \text{ ft})} = 1.4 \text{ gpm/sq ft}$

5. $\dfrac{2975 \text{ gpm}}{875 \text{ sq ft}} = 3.4 \text{ gpm/sq ft}$

PRACTICE PROBLEMS 4.4

1. $\dfrac{4950 \text{ gpm}}{(15 \text{ ft})(15 \text{ ft})} = 22 \text{ gpm/sq ft}$

2. $\dfrac{5100 \text{ gpm}}{(25 \text{ ft})(15 \text{ ft})} = 14 \text{ gpm/sq ft}$

3. $\dfrac{3300 \text{ gpm}}{(20 \text{ ft})(15 \text{ ft})} = 11 \text{ gpm/sq ft}$

4. $\dfrac{3200 \text{ gpm}}{(20 \text{ ft})(30 \text{ ft})} = 5.3 \text{ gpm/sq ft}$

5. $\dfrac{3800 \text{ gpm}}{(20 \text{ ft})(20 \text{ ft})} = 9.5 \text{ gpm/sq ft}$

PRACTICE PROBLEMS 4.5

1. $\dfrac{3,890,000 \text{ gal}}{(15 \text{ ft})(40 \text{ ft})}$ = 6483 gal/sq ft

2. $\dfrac{1,680,000 \text{ gpm}}{(15 \text{ ft})(15 \text{ ft})}$ = 7467 gal/sq ft

3. $\dfrac{3,960,000 \text{ gal}}{(20 \text{ ft})(25 \text{ ft})}$ = 7920 gal/sq ft

4. $\dfrac{1,339,200 \text{ gal}}{(15 \text{ ft})(12 \text{ ft})}$ = 7440 gal/sq ft

5. $\dfrac{5,625,000 \text{ gal}}{(30 \text{ ft})(25 \text{ ft})}$ = 7500 gal/sq ft

PRACTICE PROBLEMS 4.6

1. $\dfrac{1,397,000 \text{ gpd}}{157 \text{ ft}}$ = 8,898 gpd/ft

2. $\dfrac{2,320,000 \text{ gpd}}{(3.14)(60 \text{ ft})}$ = 12,314 gpd/ft

3. $\dfrac{2,600,000 \text{ gpd}}{235 \text{ ft}}$ = 11,064 gpd/ft

4. $\dfrac{(1200 \text{ gpm})(1440 \text{ min/day})}{(3.14)(70 \text{ ft})}$ = 7862 gpd/ft

5. $\dfrac{2819 \text{ gpm}}{188 \text{ ft}}$ = 15 gpm/ft

PRACTICE PROBLEMS 4.7

1. $\dfrac{(215 \text{ mg}/L)(2.24 \text{ MGD})(8.34 \text{ lbs/gal})}{25.1 \quad 1000\text{-cu ft}}$ = 160 lbs BOD/day/1000 cu ft

2. $\dfrac{(175 \text{ mg}/L)(0.122 \text{ MGD})(8.34 \text{ lbs/gal})}{3.7 \text{ ac}}$ = 48 lbs BOD/day/ac

3. $\dfrac{(125 \text{ mg}/L)(2.96 \text{ MGD})(8.34 \text{ lbs/gal})}{34 \quad 1000\text{-cu ft}}$ = 91 lbs BOD/day/1000 cu ft

4. $\dfrac{(130 \text{ mg/}L)(2.15 \text{ MGD})(8.34 \text{ lbs/gal})}{800 \text{ 1000-sq ft}} = 2.9$ lbs BOD/day/1000 sq ft

5. $\dfrac{(140 \text{ mg/}L)(3.4 \text{ MGD})(8.34 \text{ lbs/gal})}{25.4 \text{ 1000-cu ft}} = 156$ lbs BOD/day/1000 cu ft

PRACTICE PROBLEMS 4.8

1. $\dfrac{(207 \text{ mg/}L)(3.35 \text{ MGD})(8.34 \text{ lbs/gal})}{(1950 \text{ mg/}L)(0.4 \text{ MG})(8.34 \text{ lbs/gal})} = 0.9$

2. $\dfrac{(192 \text{ mg/}L)(3.15 \text{ MGD})(8.34 \text{ lbs/gal})}{(1690 \text{ mg/}L)(0.27 \text{ MG})(8.34 \text{ lbs/gal})} = 1.3$

3. $\dfrac{(147 \text{ mg/}L)(2.26 \text{ MGD})(8.34 \text{ lbs/gal})}{x \text{ lbs MLVSS}} = 0.7$

   $x = 3958$ lbs MLVSS

4. $\dfrac{(155 \text{ mg/}L)(1.92 \text{ MGD})(8.34 \text{ lbs/gal})}{(1880 \text{ mg/}L)(0.245 \text{ MG})(8.34 \text{ lbs/gal})} = 0.6$

5. $\dfrac{(185 \text{ mg/}L)(2.94 \text{ MGD})(8.34 \text{ lbs/gal})}{(x \text{ mg/}L)(0.45 \text{ MG})(8.34 \text{ lbs/gal}} = 0.5$

   $x = 2417$ mg/L MLVSS

PRACTICE PROBLEMS 4.9

1. $\dfrac{(2780 \text{ mg/}L)(3.75 \text{ MGD})(8.34 \text{ lbs/gal})}{(0.785)(80 \text{ ft})(80 \text{ ft})} = 17.3$ lbs MLSS/day/sq ft

2. $\dfrac{(2950 \text{ mg/}L)(4.07 \text{ MGD})(8.34 \text{ lbs/gal})}{(0.785)(75 \text{ ft})(75 \text{ ft})} = 22.7$ lbs MLSS/day/sq ft

3. $\dfrac{(x \text{ mg/}L)(3.59 \text{ MGD})(8.34 \text{ lbs/gal})}{(0.785)(60 \text{ ft})(60 \text{ ft})} = 28$ lbs MLSS/day/sq ft

   $x = 2643$ mg/L MLSS

4. $\dfrac{(2180 \text{ mg/}L)(3.2 \text{ MGD})(8.34 \text{ lbs/gal})}{(0.785)(50 \text{ ft})(50 \text{ ft})} = 29.6$ lbs MLSS/day/sq ft

PRACTICE PROBLEMS 4.9—Cont'd

5. $\dfrac{(x \text{ mg}/L)(2.96 \text{ MGD})(8.34 \text{ lbs/gal})}{(0.785)(55 \text{ ft})(55 \text{ ft})} = 20$ lbs MLSS/day/sq ft

   $x = 1924$ mg/L MLSS

PRACTICE PROBLEMS 4.10

1. $\dfrac{11,650 \text{ lbs VS/day}}{32,600 \text{ cu ft}} = 0.36$ lbs VS/day/cu ft

2. $\dfrac{(122,000 \text{ lbs/day})(0.065)(0.70)}{(0.785)(50 \text{ ft})(50 \text{ ft})(20 \text{ ft})} = 0.14$ lbs VS/day/cu ft

3. $\dfrac{(139,000 \text{ lbs/day})(0.06)(0.69)}{(0.785)(45 \text{ ft})(45 \text{ ft})(19 \text{ ft})} = 0.19$ lbs VS/day/cu ft

4. $\dfrac{(19,200 \text{ gpd})(8.34 \text{ lbs/gal})(0.053)(0.68)}{(0.785)(35 \text{ ft})(35 \text{ ft})(15 \text{ ft})} = 0.40$ lbs VS/day/cu ft

5. $\dfrac{(21,000 \text{ gpd})(8.8 \text{ lbs/gal})(0.052)(0.71)}{(0.785)(40 \text{ ft})(40 \text{ ft})(19 \text{ ft})} = 0.29$ lbs VS/day/cu ft

PRACTICE PROBLEMS 4.11

1. $\dfrac{1930 \text{ lbs VS Added/day}}{31,200 \text{ lbs VS}} = 0.06$

2. $\dfrac{550 \text{ lbs VS Added/day}}{(172,700 \text{ lbs})(0.058)(0.65)} = 0.08$

3. $\dfrac{(61,200 \text{ lbs/day})(0.054)(0.71)}{(110,000 \text{ gal})(8.34 \text{ lbs/gal})(0.065)(0.57)} = 0.07$

4. $\dfrac{x \text{ lbs VS Added/day}}{(108,000 \text{ gal})(8.34 \text{ lbs/gal})(0.058)(0.56)} = 0.08$

   $x = 2340$ lbs/day VS

5. $\dfrac{(7,700 \text{ gpd})(8.34 \text{ lbs/day})(0.044)(0.72)}{x \text{ lbs VS}} = 0.06$

   $x = 33,907$ lbs VS

PRACTICE PROBLEMS 4.12

1. 1530 people/4.7 ac = 326 people/ac

2. 3825 people/9 ac = 425 people/ac

3. $\dfrac{(1710 \text{ mg}/L)(0.372 \text{ MGD})(8.34 \text{ lbs/gal})}{0.2 \text{ lbs/day}}$ = 26,526 people

4. $\dfrac{5000 \text{ people}}{x \text{ ac}}$ = 350 people/ac

   x = 14.3 ac

5. $\dfrac{(2190 \text{ mg}/L)(0.098 \text{ MGD})(8.34 \text{ lbs/gal})}{0.2 \text{ lbs/day}}$ = 8950 people

CHAPTER 4 ACHIEVEMENT TEST

1. $\dfrac{2,140,000 \text{ gpd}}{(0.785)(75 \text{ ft})(75 \text{ ft})}$ = 485 gpd/sq ft

2. $\dfrac{2940 \text{ gpm}}{180 \text{ sq ft}}$ = 16.3 gpm/sq ft

3. $\dfrac{1,990,000 \text{ gpd}}{(3.14)(75 \text{ ft})}$ = 8450 gpd/ft

4. $\dfrac{3,100,000 \text{ gpd}}{(0.785)(80 \text{ ft})(80 \text{ ft})}$ = 617 gpd/sq ft

5. $\dfrac{(156 \text{ mg}/L)(1.8 \text{ MGD})(8.34 \text{ lbs/gal})}{x \text{ lbs MLVSS}}$ = 0.6

   x = 3903 lbs MLVSS

6. $\dfrac{400 \text{ lbs VS Added/day}}{(175,000 \text{ lbs})(0.062)(0.68)}$ = 0.05

7. $\dfrac{(2640 \text{ mg}/L)(3.45 \text{ MGD})(8.34 \text{ lbs/gal})}{(0.785)(75 \text{ ft})(75 \text{ ft})}$ = 17 lbs/day/sq ft

8. $\dfrac{(110,000 \text{ lbs/day})(0.068)(0.70)}{(0.785)(60 \text{ ft})(60 \text{ ft})(22 \text{ ft})}$ = 0.08

CHAPTER 4 ACHIEVEMENT TEST—Cont'd

9. $\dfrac{3.95 \text{ ac-ft/day}}{20 \text{ ac}} = 0.2 \text{ ft/day}$

   Then $(0.2 \text{ ft/day})(12 \text{ in./ft}) = 2.4 \text{ in./day}$

10. $\dfrac{(172 \text{ mg/}L)(3.154 \text{ MGD})(8.34 \text{ lbs/gal})}{(x \text{ mg/}L)(0.377 \text{ MG})(8.34 \text{ lbs/gal})} = 0.5$

    $x = 2878 \text{ mg/}L \text{ MLVSS}$

11. $\dfrac{(3180 \text{ mg/}L)(4.1 \text{ MGD})(8.34 \text{ lbs/gal})}{(0.785)(70 \text{ ft})(70 \text{ ft})} = 28 \text{ lbs/day/sq ft}$

    $x = 2878 \text{ mg/}L \text{ MLVSS}$

12. $\dfrac{2{,}100{,}000 \text{ gpd}}{(90 \text{ ft})(30 \text{ ft})} = 778 \text{ gpd/sq ft}$

13. $\dfrac{1{,}740{,}000 \text{ gal}}{(20 \text{ ft})(15 \text{ ft})} = 5800 \text{ gal/sq ft}$

14. $\dfrac{(140 \text{ mg/}L)(2.68 \text{ MGD})(8.34 \text{ lbs/gal})}{(1890 \text{ mg/}L)(0.29 \text{ MG})(8.34 \text{ lbs/gal})} = 0.7$

15. $\dfrac{x \text{ lbs VS Added/day}}{(23{,}800 \text{ gal})(8.34 \text{ lbs/gal})(0.057)(0.58)} = 0.07$

    $x = 459 \text{ lbs/day}$

16. $\dfrac{3000 \text{ gpm}}{(40 \text{ ft})(25 \text{ ft})} = 3 \text{ gpm/sq ft}$

17. $\dfrac{(110 \text{ mg/}L)(3.1 \text{ MGD})(8.34 \text{ lbs/gal})}{20.1 \quad 1000\text{-cu ft}} = 141 \text{ lbs BOD/day/1000 cu ft}$

18. $\dfrac{2{,}460{,}000 \text{ gpd}}{(3.14)(70 \text{ ft})} = 11{,}192 \text{ gpd/ft}$

19. $1800 \text{ people}/5.2 \text{ ac} = 346 \text{ people/ac}$

20. $\dfrac{(120 \text{ mg/}L)(2.41 \text{ MGD})(8.34 \text{ lbs/gal})}{750 \quad 1000\text{-sq ft}} = 3.2 \text{ lbs BOD/day/1000 sq ft}$

21. $\dfrac{1889 \text{ gpm}}{(40 \text{ ft})(20 \text{ ft})} = 2.4 \text{ gpm/sq ft}$

Chapter 5

PRACTICE PROBLEMS 5.1

1. $\dfrac{(30 \text{ ft})(15 \text{ ft})(7 \text{ ft})(7.48 \text{ gal/cu ft})}{937.5 \text{ gpm}} = 25 \text{ min}$

2. $\dfrac{(75 \text{ ft})(25 \text{ ft})(10 \text{ ft})(7.48 \text{ gal/cu ft})}{66,667 \text{ gph}} = 2.1 \text{ hrs}$

3. $\dfrac{(4 \text{ ft})(5 \text{ ft})(2.5 \text{ ft})(7.48 \text{ gal/cu ft})}{(5 \text{ gpm})(60 \text{ min/hr})} = 1.2 \text{ hrs}$

4. $\dfrac{(0.785)(70 \text{ ft})(70 \text{ ft})(12 \text{ ft})(7.48 \text{ gal/cu ft})}{202,917 \text{ gph}} = 1.7 \text{ hrs}$

5. $\dfrac{(400 \text{ ft})(550 \text{ ft})(5 \text{ ft})(7.48 \text{ gal/cu ft})}{219,400 \text{ gpd}} = 37.5 \text{ days}$

PRACTICE PROBLEMS 5.2

1. $\dfrac{11,900 \text{ lbs MLSS}}{2627 \text{ lbs/day SS}} = 4.5 \text{ days}$

2. $\dfrac{(2670 \text{ mg/}L \text{ MLSS})(0.28 \text{ MG})(8.34 \text{ lbs/gal})}{(128 \text{ mg/}L)(0.958 \text{ MGD})(8.34 \text{ lbs/gal})} = 6.1 \text{ days}$

3. $\dfrac{(2960 \text{ mg/}L \text{ MLSS})(0.22 \text{ MG})(8.34 \text{ lbs/gal})}{(84 \text{ mg/}L)(1.46 \text{ MGD})(8.34 \text{ lbs/gal})} = 5.3 \text{ days}$

4. $\dfrac{(x \text{ mg/}L \text{ MLSS})(0.195 \text{ MG})(8.34 \text{ lbs/gal})}{(70 \text{ mg/}L)(1.25 \text{ MGD})(8.34 \text{ lbs/gal})} = 6.5 \text{ days}$

   $x = 2917 \text{ mg/}L \text{ MLSS}$

5. $\dfrac{x \text{ lbs MLSS}}{1560 \text{ lbs/day SS}} = 5.5 \text{ days}$

   $x = 8580 \text{ lbs MLSS}$

PRACTICE PROBLEMS 5.3

1. $\dfrac{(3100 \text{ mg}/L)(0.46 \text{ MG})(8.34 \text{ lbs/gal})}{1540 \text{ lbs/day Wasted} + 330 \text{ lbs/day in SE}} = 6.4 \text{ days}$

2. $\dfrac{(2810 \text{ mg}/L \text{ MLSS})(0.325 \text{ MG})(8.34 \text{ lbs/gal})}{(5340 \text{ mg}/L \text{ SS})(0.0185 \text{ MG})(8.34 \text{ lbs/gal}) + (15 \text{ mg}/L \text{ SS})(2.15 \text{ MGD})(8.34 \text{ lbs/gal})}$

   $= \dfrac{7617 \text{ lbs}}{824 \text{ lbs/day} + 269 \text{ lbs/day}} = 7.0 \text{ days}$

3. $\dfrac{(2440 \text{ mg}/L \text{ MLSS})(1.5 \text{ MG})(8.34 \text{ lbs/gal})}{(6120 \text{ mg}/L \text{ SS})(0.075 \text{ MGD})(8.34 \text{ lbs/gal}) + (18 \text{ mg}/L)(2.6 \text{ MGD})(8.34 \text{ lbs/gal})}$

   $= \dfrac{30{,}524 \text{ lbs MLSS}}{3828 \text{ lbs/day} + 390 \text{ lbs/day}} = 8.0 \text{ days}$

4. $\dfrac{(x \text{ mg}/L)(0.755 \text{ MG})(8.34 \text{ lbs/gal})}{(6320 \text{ mg}/L)(0.030 \text{ MGD})(8.34 \text{ lbs/gal}) + (22 \text{ mg}/L)(2.4 \text{ MGD})(8.34 \text{ lbs/gal})} = 8 \text{ days}$

   First simplify:

   $\dfrac{(x \text{ mg}/L)(0.755 \text{ MG})(8.34 \text{ lbs/gal})}{1581 \text{ lbs/day} + 440 \text{ lbs/day}} = 8 \text{ days}$

   $\dfrac{(x \text{ mg}/L)(0.755 \text{ MG})(8.34 \text{ lbs/gal})}{2021 \text{ lbs/day}} = 8 \text{ days}$

   Now solve for the unknown value:

   $x = 2568 \text{ mg}/L \text{ MLSS}$

CHAPTER 5 ACHIEVEMENT TEST

1. $\dfrac{(70 \text{ ft})(25 \text{ ft})(12 \text{ ft})(7.48 \text{ gal/cu ft})}{65{,}833 \text{ gph}} = 2.4 \text{ hrs}$

2. $\dfrac{12{,}400 \text{ lbs MLSS}}{2750 \text{ lbs/day SS}} = 4.5 \text{ days}$

3. $\dfrac{(2950 \text{ mg}/L \text{ MLSS})(0.47 \text{ MG})(8.34 \text{ lbs/gal})}{1620 \text{ lbs/day wasted} + 320 \text{ lbs/day in SE}} = 6.0 \text{ days}$

4. $\dfrac{(35 \text{ ft})(15 \text{ ft})(8 \text{ ft})(7.48 \text{ gal/cu ft})}{1125 \text{ gpm}} = 28 \text{ min}$

5. $$\frac{(2740 \text{ mg}/L \text{ MLSS})(0.325 \text{ MG})(8.34 \text{ lbs/gal})}{(5910 \text{ mg}/L)(0.0192 \text{ MGD})(8.34 \text{ lbs/gal}) + (16 \text{ mg}/L)(2.3 \text{ MGD})(8.34 \text{ lbs/gal})}$$

$$= \frac{7427 \text{ lbs MLSS}}{946 \text{ lbs/day} + 307 \text{ lbs/day}} = 5.9 \text{ days}$$

6. $$\frac{(3140 \text{ mg}/L)(0.32 \text{ MG})(8.34 \text{ lbs/gal})}{(105 \text{ mg}/L)(2.24 \text{ MGD})(8.34 \text{ lbs/gal})} = 4.3 \text{ days}$$

7. $$\frac{(2370 \text{ mg}/L)(1.7 \text{ MG})(8.34 \text{ lbs/gal})}{(6210 \text{ mg}/L)(0.0694 \text{ MGD})(8.34 \text{ lbs/gal}) + (20 \text{ mg}/L)(2.72 \text{ MGD})(8.34 \text{ lbs/gal})}$$

$$= \frac{33,602 \text{ lbs}}{3594 \text{ lbs/day} + 454 \text{ lbs/day}} = 8.3 \text{ days}$$

8. $$\frac{(2580 \text{ mg}/L)(0.24 \text{ MG})(8.34 \text{ lbs/gal})}{(145 \text{ mg}/L)(0.81 \text{ MGD})(8.34 \text{ lbs/gal})} = 5.3 \text{ days}$$

9. $$\frac{(0.785)(5 \text{ ft})(5 \text{ ft})(3 \text{ ft})(7.48 \text{ gal/cu ft})}{10 \text{ gpm}} = 44 \text{ min}$$

10. $$\frac{x \text{ lbs MLSS}}{(135 \text{ mg}/L)(1.38 \text{ MGD})(8.34 \text{ lbs/gal})} = 5 \text{ days}$$

$$x = 7769 \text{ lbs MLSS}$$

11. $$\frac{(300 \text{ ft})(480 \text{ ft})(5 \text{ ft})(7.48 \text{ gal/cu ft})}{195,000 \text{ gpd}} = 27.6 \text{ days}$$

12. $$\frac{(x \text{ mg}/L \text{ MLSS})(0.71 \text{ MG})(8.34 \text{ lbs/gal})}{(6240 \text{ mg}/L)(0.036 \text{ MGD})(8.34 \text{ lbs/gal}) + (15 \text{ mg}/L)(2.83 \text{ MGD})(8.34 \text{ lbs/gal})} = 8 \text{ days}$$

First simplify:

$$\frac{(x \text{ mg}/L)(0.71 \text{ MG})(8.34 \text{ lbs/gal})}{1873 \text{ lbs/day} + 354 \text{ lbs/day}} = 8 \text{ days}$$

or $$\frac{(x \text{ mg}/L)(0.71 \text{ MG})(8.34 \text{ lbs/gal})}{2227 \text{ lbs/day}} = 8 \text{ days}$$

Then solve for x:

$$x = 3008 \text{ mg}/L \text{ MLSS}$$

Chapter 6

PRACTICE PROBLEMS 6.1

1. $\dfrac{98 \text{ mg/}L \text{ Rem.}}{120 \text{ mg/}L} \times 100 = 82\%$

2. $\dfrac{230 \text{ mg/}L \text{ Rem.}}{245 \text{ mg/}L} \times 100 = 94\%$

3. $\dfrac{208 \text{ mg/}L \text{ Rem.}}{270 \text{ mg/}L} \times 100 = 77\%$

4. $\dfrac{95 \text{ mg/}L \text{ Rem.}}{230 \text{ mg/}L} \times 100 = 41\%$

5. $\dfrac{168 \text{ mg/}L \text{ Rem.}}{305 \text{ mg/}L} \times 100 = 55\%$

PRACTICE PROBLEMS 6.2

1. $5.8 = \dfrac{x \text{ lbs/day Solids}}{(3610 \text{ gal})(8.34 \text{ lbs/gal})} \times 100$

   $x = 1746 \text{ lbs/day Solids}$

2. $\dfrac{0.68 \text{ grams Solids}}{13.05 \text{ grams Sludge}} \times 100 = 5.2\% \text{ solids}$

3. $\dfrac{1430 \text{ lbs/day Solids}}{x \text{ lbs/day Sludge}} \times 100 = 3.2$

   $x = 46,667 \text{ lbs/day}$

4. $4.4 = \dfrac{265 \text{ lbs/day SS}}{(x \text{ gpd})(8.34 \text{ lbs/gal})} \times 100$

   $x = 722 \text{ gpd}$

5. $3.4 = \dfrac{x \text{ lbs/day Solids}}{286,000 \text{ lbs/day Sludge}} \times 100$

   $x = 9724 \text{ lbs/day Solids}$

PRACTICE PROBLEMS 6.3

1. $$\frac{\dfrac{(3000 \text{ gpd})(8.34 \text{ lbs/gal})(4.5)}{100} + \dfrac{(4200 \text{ gpd})(8.34 \text{ lbs/gal})(3.8)}{100}}{(3000 \text{ gpd})(8.34 \text{ lbs/gal}) + (4200 \text{ gpd})(8.34 \text{ lbs/gal})} \times 100$$

First simplify:

$$\frac{1126 \text{ lbs/day Sol.} + 1331 \text{ lbs/day Sol.}}{25{,}020 \text{ lbs/day} + 35{,}028 \text{ lbs/day sludge}} \times 100$$
$$\text{sludge}$$

$$= \frac{2457 \text{ lbs/day Solids}}{60{,}048 \text{ lbs/day Sludge}} \times 100$$

$$= 4.1\%$$

2. $$\frac{\dfrac{(8200 \text{ gpd})(8.34 \text{ lbs/gal})(5.2)}{100} + \dfrac{(7000 \text{ gpd})(8.34 \text{ lbs/gal})(4.2)}{100}}{(8200 \text{ gpd})(8.34 \text{ lbs/gal}) + (7000 \text{ gpd})(8.34 \text{ lbs/gal})} \times 100$$

Simplify:

$$\frac{3556 \text{ lbs/day Sol.} + 2452 \text{ lbs/day Sol.}}{68{,}388 \text{ lbs/day Sludge} + 58{,}380 \text{ lbs/day Sludge}} \times 100$$

$$= \frac{6008 \text{ lbs/day Solids}}{126{,}768 \text{ lbs/day Sludge}} \times 100$$

$$= 4.7\%$$

3. $$\frac{\dfrac{(4840 \text{ gpd})(8.34)(4.9)}{100} + \dfrac{(5200 \text{ gpd})(8.34)(3.6)}{100}}{(4840)(8.34) + (5200)(8.34)} \times 100$$

$$= \frac{1978 \text{ lbs/day} + 1561 \text{ lbs/day}}{40{,}366 \text{ lbs/day} + 43{,}368 \text{ lbs/day}} \times 100$$

$$= \frac{3539 \text{ lbs/day Solids}}{83{,}734 \text{ lbs/day Sludge}} \times 100$$

$$= 4.2\%$$

PRACTICE PROBLEMS 6.3—Cont'd

4.
$$\frac{(9010 \text{ gpd})(8.34)\frac{(4.2)}{100} + (10,760 \text{ gpd})(8.34)\frac{(6.5)}{100}}{(9010)(8.34) + (10,760)(8.34)} \times 100$$

$$= \frac{3156 \text{ lbs/day} + 5833 \text{ lbs/day}}{75,143 \text{ lbs/day} + 89,738 \text{ lbs/day}} \times 100$$

$$= \frac{8989 \text{ lbs/day}}{164,881 \text{ lbs/day}} \times 100$$

$$= 5.5\%$$

PRACTICE PROBLEMS 6.4

1. (3340 lbs/day Solids)(0.70) = 2338 lbs/day VS

2. (4070 gpd Sludge)(8.34 lbs/gal)(0.07)(0.72) = 1711 lbs/day VS

3. (6400 gpd Sludge)(8.34 lbs/gal)(0.065)(0.67) = 2325 lbs/day VS

4. (24,510 lbs/day Sludge)(0.068)(0.66) = 1100 lbs/day VS

5. (2530 gpd Sludge)(8.34 lbs/gal)(0.062)(0.71) = 929 lbs/day VS

PRACTICE PROBLEMS 6.5

1. $18 = \dfrac{x \text{ gal Seed Sludge}}{280,000 \text{ gal Volume}} \times 100$

   $x = 50,400$ gal Seed Sludge

2. $20 = \dfrac{x \text{ gal Seed Sludge}}{(0.785)(70 \text{ ft})(70 \text{ ft})(21 \text{ ft})(7.48 \text{ gal/cu ft})} \times 100$

   $x = 120,842$ gal Seed Sludge

3. $15 = \dfrac{x \text{ gal Seed Sludge}}{(0.785)(50 \text{ ft})(50 \text{ ft})(21 \text{ ft})(7.48 \text{ gal/cu ft})} \times 100$

   $x = 46,240$ gal Seed Sludge

4. $x = \dfrac{87,500 \text{ gal}}{(0.785)(60 \text{ ft})(60 \text{ ft})(23 \text{ ft})(7.48 \text{ gal/cu ft})} \times 100$

   $x = 18\%$

PRACTICE PROBLEMS 6.6

1. $x = \dfrac{2 \text{ lbs Chemical}}{80 \text{ lbs Solution}} \times 100$

   $x = 2.5\%$ Strength

2. (6 oz Polymer = 0.4 lbs Polymer)

   $x = \dfrac{0.4 \text{ lbs Polymer}}{(8 \text{ gal})(8.34 \text{ lbs/gal}) + 0.4 \text{ lbs}} \times 100$

   $x = \dfrac{0.4}{66.7 \text{ lbs} + 0.4 \text{ lbs}} \times 100$

   $x = 0.6\%$ Strength

3. $1.5 = \dfrac{x \text{ lbs Polymer}}{(50 \text{ gal})(8.34 \text{ lbs/gal}) + x \text{ lbs Polymer}} \times 100$

   $0.015 = \dfrac{x}{417 + x} \times 100$

   $(0.015)(417 + x) = x$

   $6.255 + 0.015\,x = x$

   $6.255 = x - 0.015x$

   $6.255 = 0.985x$

   $6.4 \text{ lbs} = x$

4. (500 g = 1.1 lbs dry polymer)

   $x = \dfrac{1.1 \text{ lbs Polymer}}{(8 \text{ gal})(8.34 \text{ lbs/gal}) + 1.1 \text{ lbs}} \times 100$

   $x = \dfrac{1.1}{66.7 + 1.1} \times 100$

   $x = 1.6\%$

PRACTICE PROBLEMS 6.6—Cont'd

5. First calculate lbs of chemical required:

$$1.8 = \frac{x \text{ lbs Chemical}}{(5 \text{ gal})(8.34 \text{ lbs/gal}) + x \text{ lbs}}$$

$$\frac{1.8}{100} = \frac{x \text{ lbs}}{41.7 \text{ lbs} + x \text{ lbs}}$$

$$0.018 = \frac{x}{41.7 \text{ lbs} + x \text{ lbs}}$$

$$(0.018)(41.7 + x) = x$$

$$0.75 + 0.018\,x = x$$

$$0.75 = x - 0.018\,x$$

$$0.75 = 0.982\,x$$

$$0.76 \text{ lbs} = x$$

Then determine grams chem. required:

$$(0.76 \text{ lbs})(454 \text{ grams/lb}) = 345 \text{ grams}$$

PRACTICE PROBLEMS 6.7

1.
$$x = \frac{(15 \text{ lbs Sol'n 1})\left(\frac{9}{100}\right) + (100 \text{ lbs Sol'n 2})\left(\frac{1}{100}\right)}{15 \text{ lbs} + 100 \text{ lbs}} \times 100$$

$$x = \frac{(15)(0.09) + (100)(0.01)}{115} \times 100$$

$$x = \frac{1.35 + 1}{115} \times 100$$

$$x = 2\% \text{ Strength}$$

2.
$$x = \frac{(10 \text{ lbs})\left(\frac{12}{100}\right) + (330 \text{ lbs})\left(\frac{0.3}{100}\right)}{10 \text{ lbs} + 330 \text{ lbs}} \times 100$$

$$x = \frac{(10)(0.12) + (330)(0.003)}{340} \times 100$$

$$x = \frac{1.2 + 0.99}{340} \times 100$$

$$x = 0.6\% \text{ Strength}$$

3.

$$x = \frac{\dfrac{(20\ lbs)(12)}{100} + \dfrac{(445\ lbs)(0.5)}{100}}{20\ lbs + 445\ lbs} \times 100$$

$$x = \frac{(10)(0.12) + (330)(0.003)}{340} \times 100$$

$$x = \frac{2.4\ lbs + 2.2\ lbs}{465\ lbs} \times 100$$

$x = 1.0\%$ Strength

4. First determine lbs of each solution:

11% solution: $(10\ gal)(9.8\ lbs/gal) = 98\ lbs$

0.1% solution: $(60\ gal)(8.34\ lbs/gal) = 500\ lbs$

Then continue as with problems 1-3:

$$x = \frac{\dfrac{(98\ lbs)(11)}{100} + \dfrac{(500\ lbs)(0.1)}{100}}{98\ lbs + 500\ lbs} \times 100$$

$$x = \frac{10.8\ lbs + 0.5\ lbs}{598\ lbs} \times 100$$

$x = 1.9\%$ Strength

PRACTICE PROBLEMS 6.8

1. $x = \dfrac{17\ hp}{22} \times 100$

   $x = 77\%$

2. $x = \dfrac{43\ hp}{50\ hp} \times 100$

   $x = 86\%$

3. $89 = \dfrac{x}{25} \times 100$

   $x = 22\ hp$

4. $62 = \dfrac{x}{50} \times 100$

   $x = 31\ hp$

PRACTICE PROBLEMS 6.8—Cont'd

5. $86 = \dfrac{34.4\ hp}{x} \times 100$

   $x = 40\ hp$

CHAPTER 6 ACHIEVEMENT TEST

1. $x = \dfrac{209}{230} \times 100$

   $x = 91\%$

2. $(7200\ gpd)(8.34\ lbs/gal)(0.07)(0.68) = 2858\ lbs/day\ VS$

3. $x = \dfrac{\dfrac{(2600\ gpd)(8.34\ lbs/gal)(4.2)}{100} + \dfrac{(3380\ gpd)(8.34\ lb/gal)(3.5)}{100}}{(2600\ gpd)(8.34\ lbs/gal) + (3380\ gpd)(8.34\ lb/gal)} \times 100$

   $x = \dfrac{911\ lbs/day + 987\ lbs/day}{21{,}684\ lbs/day + 28{,}189\ lbs/day} \times 100$

   $x = \dfrac{1898\ lbs/day}{49{,}873\ lbs/day} \times 100$

   $x = 4\%$

4. $x = \dfrac{3\ lbs\ Chemical}{92\ lbs\ Solution} \times 100$

   $x = 3\%$

5. $20 = \dfrac{x\ gal}{(0.785)(80\ ft)(80\ ft)(23\ ft)(7.48\ gal/cu\ ft)} \times 100$

   $x = 172{,}866\ gal$

6. $x = \dfrac{\dfrac{(12\ lbs)(10)}{100} + \dfrac{(400\ lbs)(0.15)}{100}}{12\ lbs + 400\ lbs} \times 100$

   $x = \dfrac{1.2\ lbs + 0.6\ lbs}{412\ lbs} \times 100$

   $x = 0.4\%$

7. $x = \dfrac{20.4 \text{ hp}}{25.5 \text{ hp}} \times 100$

    $x = 80\% \text{ Effic.}$

8. $x = \dfrac{269 \text{ mg/}L}{350 \text{ mg/}L} \times 100$

    $x = 77\% \text{ Rem. Effic.}$

9. $x = \dfrac{\dfrac{(8700 \text{ gpd})(8.34 \text{ lbs/gal})(4.8)}{100} + \dfrac{(11{,}200 \text{ gpd})(8.34 \text{ lb/gal})(6.2)}{100}}{(8700 \text{ gpd})(8.34 \text{ lbs/gal}) + (11{,}200 \text{ gpd})(8.34 \text{ lb/gal})} \times 100$

    $x = \dfrac{3483 \text{ lbs/day} + 5791 \text{ lbs/day}}{72{,}558 \text{ lbs/day} + 93{,}408 \text{ lbs/day}} \times 100$

    $x = \dfrac{9274 \text{ lbs/day}}{165{,}966 \text{ lbs/day}} \times 100$

    $x = 5.6\%$

10. $x = (7400 \text{ gpd})(8.34 \text{ lbs/gal})\dfrac{(6.2)}{100}\dfrac{(72)}{100}$

     $= 2755 \text{ lbs/day VS}$

11. $5.7 = \dfrac{x \text{ lbs/day solids}}{35{,}400 \text{ lbs/day sludge}} \times 100$

     $x = 2018 \text{ lbs/day}$

12. $0.5 = \dfrac{x \text{ lbs Polymer}}{(75 \text{ gal})(8.34 \text{ lbs/gal}) + x \text{ lbs}} \times 100$

     $0.5 = \dfrac{x \text{ lbs}}{625.5 \text{ lbs} + x \text{ lbs}} \times 100$

     $(0.5)(625.5 + x) = x$

     $312.75 + 0.5x = x$

     $312.75 = x - 0.5x$

     $312.75 = 0.5x$

     $626 \text{ lbs} = x$

CHAPTER 6 ACHIEVEMENT TEST—Cont'd

13. $25 = \dfrac{x \text{ gal}}{190,000 \text{ gal}} \times 100$

   $x = 47,500 \text{ gal}$

14. $x = (6800 \text{ lbs/day solids})(0.71)$

   $x = 4828 \text{ lbs/day VS}$

15. $x = \dfrac{(8 \text{ lbs})\left(\dfrac{9}{100}\right) + (90 \text{ lbs})\left(\dfrac{1}{100}\right)}{8 \text{ lbs} + 90 \text{ lbs}} \times 100$

   $x = \dfrac{0.7 \text{ lbs} + 0.9 \text{ lbs}}{98 \text{ lbs}} \times 100$

   $x = 1.6\%$

16. $x = \dfrac{38 \text{ hp}}{50 \text{ hp}} \times 100$

   $x = 76\% \text{ Effic.}$

17. $67 = \dfrac{x \text{ whp}}{75 \text{ mhp}} \times 100$

   $x = 50 \text{ whp}$

Chapter 7

PRACTICE PROBLEMS 7.1

1. 8.9 lbs/1 gal = 8.9 lbs/gal

2. $\dfrac{67.1 \text{ lbs/cu ft}}{7.48 \text{ gal/cu ft}} = 9.0 \text{ lbs/gal}$

3. $\dfrac{55 \text{ lbs/cu ft}}{62.4 \text{ lbs/cu ft}} = 0.9$

4. (1.3)(8.34 lbs/gal) = 10.8 lbs/gal

5. $\dfrac{8.9 \text{ lbs/gal}}{8.34 \text{ lbs/gal}} = 1.1$

PRACTICE PROBLEMS 7.2

1. $\dfrac{75 \text{ lbs}}{(30 \text{ in.})(18 \text{ in.})} = 0.1 \text{ lbs/sq in.}$

2. $(2.6 \text{ lbs/sq in.})(4) = 10.4 \text{ lbs/sq in.}$

3. $\dfrac{210 \text{ lbs}}{(3 \text{ ft})(3 \text{ ft})} = 23 \text{ lbs/sq ft};$ $\dfrac{210 \text{ lbs}}{(3 \text{ ft})(1.25 \text{ ft})} = 56 \text{ lbs/sq ft}$

4. $(6 \text{ ft})(62.4 \text{ lbs/cu ft}) = 374 \text{ lbs/sq ft}$

   An alternate method: $\dfrac{6 \text{ ft}}{2.31 \text{ ft/psi}} = 2.597 \text{ psi}$

   $(2.597 \text{ psi}) \dfrac{(144 \text{ sq in.})}{\text{sq ft}} = 374 \text{ lbs/sq ft}$

5. $\dfrac{8 \text{ ft}}{2.31 \text{ ft/psi}} = 3.5 \text{ psi}$

6. First calculate the psi for water:

   $\dfrac{4.5 \text{ ft}}{2.31 \text{ ft/psi}} = 1.9 \text{ psi}$

   Then determine psi for a different specific gravity:

   $(1.9 \text{ psi})(1.4) = 2.7 \text{ psi}$

7. First determine the pressure at 10 ft depth:

   $\dfrac{10 \text{ ft}}{2.31 \text{ ft/psi}} = 4.3 \text{ psi}$

   Then calculate total force:

   $(4.3 \text{ psi})(240 \text{ in.})(144 \text{ in.}) = 148,608 \text{ lbs}$

8. First determine the pressure in lbs/sq ft at the average depth:

   $(5)(62.4 \text{ lbs/sq ft}) = 312 \text{ lbs/sq ft}$

   Then calculate total force on the wall:

   $(312 \text{ lbs/sq ft})(25 \text{ ft})(10 \text{ ft}) = 78,000 \text{ lbs}$

PRACTICE PROBLEMS 7.2—Cont'd

9. $\dfrac{(2)(9 \text{ ft})}{3} = 6 \text{ ft}$

10. First determine the lbs/sq ft pressure at the small cylinder:

    $$\frac{50 \text{ lbs}}{(0.785)(0.67 \text{ ft})(0.67 \text{ ft})} = 142 \text{ lbs/sq ft}$$

    Then calculate total force at the large cylinder:

    $$(142 \text{ lbs/sq ft})(0.785)(2 \text{ ft})(2 \text{ ft}) = 446 \text{ lbs}$$

11. 32 psi gage + 14.7 psi atmos. = 46.7 psi absolute

PRACTICE PROBLEMS 7.3

1. 322 ft − 239 ft = 83 ft

2. Total static head is:

   790 ft − 614 ft = 176 ft

   Then TDH is:

   176 ft + 16 ft = 192 ft TDH

3. First calculate static head in psi:

   162 psi − 95 psi = 67 psi

   Then convert to ft of head:

   (67 psi)(2.31 ft/psi) = 155 ft

   And TDH is:

   155 ft + 12 ft = 167 ft TDH

4. $\dfrac{(10.11 \text{ ft head loss})(10 \text{ sections})}{100 \text{ ft section}} = 101 \text{ ft head loss of } 100 \text{ ft}$

5. There is a friction loss given for 1400 gpm and for 1500 gpm. To estimate the friction loss for 1450 gpm, use the value midway between the two values:

> Friction loss for 1400 gpm—1.87 ft/100-ft section
> Friction loss for 1500 gpm—2.13 ft/100-ft section

The friction loss value midway between these two values is:

$$\frac{1.87\ ft + 2.13\ ft}{2} = 2\ ft$$

Then determine the friction head loss:

$$\frac{(2\ ft\ head\ loss)}{100\ ft\ section}\ (20\ sections) = 40\ ft\ of\ 100\ ft$$

6. 26 ft Equivalent Length of Pipe

PRACTICE PROBLEMS 7.4

1. $Hp = \dfrac{(20\ ft)(1500\ gpm)(8.34\ lbs/gal)}{33,000\ ft\text{-}lbs/min/hp}$

   $= 7.6\ hp$

2. Brake horsepower is:

   $(20\ mhp)\dfrac{(85)}{100} = 17\ bhp$

   Water horsepower is:

   $(17\ bhp)\dfrac{(80)}{100} = 13.6\ whp$

3. $\dfrac{35\ whp}{\dfrac{85}{100}} = 41.2\ bhp$

4. $Hp = \dfrac{(60\ ft)(600\ gpm)(8.34\ lbs/gal)(1.2\ sp.\ gr.)}{33,000\ ft\text{–}lbs/min/hp}$

   $= 10.9\ whp$

5. $(45\ hp)(746\ watts/hp) = 33,570\ watts$

   Then convert watts to kilowatts:

   $\dfrac{33,570\ watts}{1000\ watts/kW} = 33.6\ kW$

PRACTICE PROBLEMS 7.4—Cont'd

6. First calculate kilowatts required:

(75 mhp)(746 watts/hp) = 55,950 watts

Then convert watts to kilowatts:

$$\frac{55,950 \text{ watts}}{1000 \text{ watts/kW}} = 55.9 \text{ kw}$$

The power consumption for the week is:

(55.9 kw)(144 hrs) = 8049.6 kWh

And the power cost is therefore:

(8049.6 kWh)($0.06125/kWh) = $493.04

PRACTICE PROBLEMS 7.5

1. $$\frac{(12 \text{ ft})(10 \text{ ft})(2.6 \text{ ft})(7.48 \text{ gal/cu ft})}{5 \text{ min}} = 467 \text{ gpm}$$

2. $$\frac{55 \text{ gallons}}{0.48 \text{ min}} = 115 \text{ gpm}$$

3. The rise in level is equivalent to a gpm pumping rate of:

$$\frac{(8 \text{ ft})(10 \text{ ft})(0.17 \text{ ft})(7.48 \text{ gal/cu ft})}{5 \text{ min}} = 20 \text{ gpm}$$

The pumping rate is therefore:

400 gpm – 20 gpm = 380 gpm

4. $$\frac{(0.75 \text{ gal})}{\text{stroke}} \frac{(30 \text{ strokes})}{\text{min}} = 22.5 \text{ gpm}$$

5. First calculate gpm pumping rate:

$$\left[(0.785)(0.67 \text{ ft})(0.67 \text{ ft})(0.33 \text{ ft})(7.48 \text{ gal/cu ft})\right]\left[40 \text{ strokes/min}\right]$$
$$= 34.8 \text{ gpm}$$

Then convert gpm pumping rate to gpd, based on total minutes of operation:

(34.8 gpm)(150 min/day) = 5220 gpd

CHAPTER 7 ACHIEVEMENT TEST

1. $\dfrac{66 \text{ lbs/cu ft}}{7.48 \text{ gal/cu ft}} = 8.82 \text{ lbs/gal}$

2. $\dfrac{(12 \text{ ft})(10 \text{ ft})(1.5 \text{ ft})(7.48 \text{ gal/cu ft})}{5 \text{ min}} = 269 \text{ gpm}$

3. $\dfrac{90 \text{ lbs}}{(36 \text{ in.})(24 \text{ in.})} = 0.1 \text{ lbs/sq in.}$

4. $\left[(0.785)(0.5 \text{ ft})(0.5 \text{ ft})(0.2 \text{ ft})(7.48 \text{ gal/cu ft})\right]\left[55 \dfrac{\text{strokes}}{\text{min}}\right]$

   $= 16 \text{ gpm}$

5. $\dfrac{(10 \text{ ft})(12 \text{ ft})(1.2 \text{ ft})(7.48 \text{ gal/cu ft})}{5 \text{ min}} = 215 \text{ gpm}$

6. $\dfrac{(70 \text{ ft})(900 \text{ gpm})(8.34 \text{ lb/gal})}{33,000 \text{ ft-lbs/min/hp}} = 15.9 \text{ hp}$

7. First calculate gpm pumping rate:

   $\left[(0.785)(0.67 \text{ ft})(0.67 \text{ ft})(0.25 \text{ ft})(7.48 \text{ gal/cu ft})\right]\left[50 \text{ strokes/min}\right]$

   $= 32.9 \text{ gpm}$

   Pumping rate is:

   $(32.9 \text{ gpm})(125 \dfrac{\text{min}}{\text{day}}) = 4113 \text{ gpd}$

8. $(1.4)(8.34 \text{ lbs/gal}) = 11.7 \text{ lbs/gal}$

9. $\dfrac{7 \text{ ft}}{2.31 \text{ ft/psi}} = 3.0 \text{ psi}$

10. Total static head is:

    $852 \text{ ft} - 760 \text{ ft} = 92 \text{ ft}$

    And TDH is:

    $92 \text{ ft} + 10 \text{ ft} = 102 \text{ ft}$

CHAPTER 7 ACHIEVEMENT TEST—Cont'd

11. 175 ft equivalent length of pipe

12. The pressure at the average depth is:

    (4.5 ft)(62.4 lbs/cu ft) = 281 lbs/sq ft

    The total force is:

    (281 lbs/sq ft)(15 ft)(9 ft) = 37,935 lbs

13. Brake horsepower is:

    $$(50 \text{ mhp})\frac{(90)}{100} = 45 \text{ bhp}$$

    Water horsepower is:

    $$(45 \text{ bhp})\frac{(85)}{100} = 38 \text{ whp}$$

14. $$\frac{(2.97 \text{ ft head loss})(30 \text{ sections})}{100 \text{ ft section}} \text{ of } 100 \text{ ft} = 89 \text{ ft head loss}$$

15. The friction head loss at 240 gpm is 0.22 ft/100-ft section. The head loss for the 1200-ft section is therefore:

    $$\frac{(0.22 \text{ ft head loss})(12 \text{ sections})}{100 \text{ ft section}} \text{ of } 100 \text{ ft} = 2.64 \text{ ft}$$

16. First calculate watts required:

    (60 mhp)(746 watts/hp) = 44,760 watts

    Then convert watts to kilowatts:

    $$\frac{44,760 \text{ watts}}{1000 \text{ watts/kW}} = 44.8 \text{ kW}$$

    The power used during the week is:

    (44.8 kW)(120 hrs) = 5376 kWh

    And the cost is:

    (5376 kWh)($0.05626/kWh) = $302.45

Chapter 8

PRACTICE PROBLEMS 8.1

1. 97 ft − 89 ft = 8 ft drawdown

2. 128 ft − 105 ft = 23 ft drawdown

3. 159 ft − 142 ft = 17 ft drawdown

4. First calculate the pumping water level:

    (3.6 psi) (2.31 ft/psi) = 8.3 ft water depth
    in sounding line

    $$\text{Pumping Water Level, ft} = \text{Length of Sounding Line, ft} - \text{Water Depth in Sounding Line, ft}$$

    = 110 ft − 8.3 ft

    = 101.7 ft

    Drawdown can now be calculated:

    101.7 ft − 82 ft = 19.7 ft

5. First calculate the static water level:

    (4.5 psi) (2.31 ft/psi) = 10.4 ft water depth
    in sounding line

    $$\text{Water Level in Well, ft} = \text{Length of Sounding Line, ft} - \text{Water Depth in Sounding Line, ft}$$

    = 150 ft 10.4 ft

    = 139.6 ft

    Drawdown can now be calculated:

    168 ft − 139.6 ft = 28.4 ft drawdown

PRACTICE PROBLEMS 8.2

1. $\dfrac{405 \text{ gallons}}{5 \text{ minutes}}$ = 81 gpm

2. $\dfrac{780 \text{ gallons}}{5 \text{ minutes}}$ = 156 gpm

PRACTICE PROBLEM 8.2—Cont'd

3. $\dfrac{835 \text{ gallons}}{5 \text{ minutes}}$ = 167 gpm

   (167 gpm) (60 min/hr) = 10,020 gph

4. $\dfrac{(0.785)\,(1\text{ ft})\,(1\text{ ft})\,(10\text{ ft})\,(7.48\text{ gal/cu ft})\,(10\text{ round trips})}{5 \text{ min}} = \dfrac{587 \text{ gal}}{5 \text{ min}}$

   $= 117 \text{ gpm}$

5. First calculate the well yield in gpm:

   $\dfrac{740 \text{ gal}}{5 \text{ min}}$ = 148 gpm

   Then convert gpm to gph flow rate:

   (148 gpm) (60 min/hr) = 8,880 gph

   If the pump operates a total of 9 hrs each day, the gallons pumped each day is:

   (8880 gph) (9 hrs/day) = 79,920 gal/day

PRACTICE PROBLEM 8.3

1. $\dfrac{190 \text{ gpm}}{26 \text{ ft}}$ = 7.3 gpm/ft

2. $\dfrac{510 \text{ gpm}}{21 \text{ ft}}$ = 24.3 gpm/ft

3. $\dfrac{1000 \text{ gpm}}{38.2 \text{ ft}}$ = 26.2 gpm/ft

4. $\dfrac{x \text{ gpm}}{41.6 \text{ ft}}$ = 31.2 gpm/ft

   x = (31.2) (41.6)

   = 1298 gpm

PRACTICE PROBLEM 8.4

1. First calculate the volume of the water-filled casing:

   (0.785) (0.5 ft) (0.5 ft) (130 ft) (7.48 gal/cu ft) = 191 gal

   Then calculate the lbs chlorine required:

   (50 mg/L) (0.000191 MG) (8.34 lbs/gal) = 0.08 lbs chlorine

2.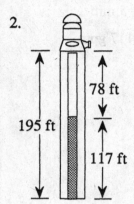

   First calculate the volume of the water-filled casing:

   (0.785) (1 ft) (1 ft) (117 ft) (7.48 gal/cu ft) = 687 gal

   Then calculate the lbs chlorine required:

   (50 mg/L) (0.000687 MG) (8.34 lbs/gal) = 0.3 lbs Chlorine

3. First determine the volume of water in the water-filled casing:

   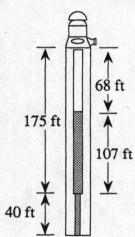

   The 12-in. diameter casing has a water-filled volume of:

   (0.785) (1 ft) (1 ft) (107 ft) (7.48 gal/cu ft) = 628 gal

   The 8-in. diameter casing has a water-filled volume of:

   (0.785) (0.67 ft) (0.67 ft) (40 ft) (7.48 gal/cu ft) = 105 gal

   The total volume of the water-filled casing is therefore:

   628 gal + 105 gal = 733 gal

   The required lbs of chlorine can now be calculated:

   (100 mg/L) (0.000733 gal) (8.34 lbs/gal) = 0.6 lbs Chlorine

4. $(x \text{ mg/L}) (0.000520 \text{ gal}) (8.34 \text{ lbs/gal}) = 0.48 \text{ lbs}$

   $$x = \frac{0.48}{(0.000520)(8.34)}$$

   $x = 111 \text{ mg/L}$

PRACTICE PROBLEM 8.4—Cont'd

5. First calculate lbs sodium hypochlorite required:

$$\frac{0.08 \text{ lbs chlorine}}{\frac{5.25}{100}} = 1.5 \text{ lbs}$$

Next, convert to gal sodium hypochlorite:

$$\frac{1.5 \text{ lbs}}{8.34 \text{ lbs/gal}} = 0.18 \text{ gal}$$

Then convert gal to fl oz:

$$(0.18 \text{ gal}) \frac{(128 \underline{\text{fl oz}})}{\text{gal}} = 23 \text{ fl oz}$$

6. First calculate the volume of the water-filled casing:

$$(0.785) (0.5 \text{ ft}) (0.5 \text{ ft}) (115 \text{ ft}) (7.48 \text{ gal/cu ft}) = 169 \text{ gal}$$

Next, determine the lbs of calcium hypochlorite required:

$$\frac{(50 \text{ mg}/L \text{ Cl}_2) (0.000169 \text{ MG}) (8.34 \text{ lbs/gal})}{\frac{65}{100}} = 0.1 \text{ calcium hypochlorite}$$

Then convert lbs to oz calcium hypochlorite:

$$(0.1 \text{ lbs}) \frac{(16 \text{ oz})}{1 \text{ lb}} = 1.6 \text{ ounces calcium hypochlorite}$$

7. The volume of the water-filled casing is:

$$(0.785) (1.5 \text{ ft}) (1.5 \text{ ft}) (95 \text{ ft}) (7.48 \text{ gal/cu ft}) = 1255 \text{ gal}$$

The lbs of chloride of lime required is therefore:

$$\frac{(100 \text{ mg}/L) (0.001255 \text{ MG}) (8.34 \text{ lbs/gal})}{\frac{25}{100}} = 4.2 \text{ lbs chloride of lime}$$

8. First calculate the lbs sodium hypochlorite required:

$$\frac{(50 \text{ mg}/L) (0.000235 \text{ MG}) (8.34 \text{ lbs/gal})}{\frac{5.25}{100}} = 1.9 \text{ lbs}$$

Next, calculate the gallons sodium hypochlorite required:

$$\frac{1.9 \text{ lbs}}{8.34 \text{ lbs/gal}} = 0.2 \text{ gal}$$

Then convert gallons sodium hypochlorite to fluid ounces:

$$(0.2 \text{ gal}) \frac{(128 \underline{\text{fl oz}})}{\text{gal}} = 25.6 \text{ fl oz sodium hypochlorite}$$

PRACTICE PROBLEM 8.5

1. (4.1 psi) (2.31 ft/psi) = 9.5 ft

2. The field head is a total of the lift <u>below</u> the discharge head centerline (pumping water level) and the discharge head <u>above</u> the discharge head centerline:

$$\text{Field Head,} \atop \text{ft} = \text{Pumping} \atop \text{Water} \atop \text{Level, ft} + \text{Discharge} \atop \text{Head, ft}$$

$$= (96 \text{ ft} + 26 \text{ ft}) + (3.8 \text{ psi}) (2.31 \text{ ft/psi})$$

$$= 122 \text{ ft} + 8.8 \text{ ft}$$

$$= 130.8 \text{ ft}$$

3. Lab Head, ft = Field Head, ft + Column Friction Loss, ft

   From the column friction loss table, the friction loss per <u>100 ft</u> for a 10-inch diameter column, 2-inch tube, and 1400 gpm flow is 1.6 ft.

   The friction loss for the entire 190-ft column is therefore:

   $$\frac{(1.6 \text{ ft loss}) (190 \text{ ft})}{100 \text{ ft}} = 3 \text{ ft}$$

   Lab head can now be determined:

   $$\text{Lab Head,} \atop \text{ft} = 135 \text{ ft} + 3 \text{ ft}$$

   $$= 138 \text{ ft}$$

4. Before calculating lab head, the column friction loss must be determined. From the table given in Appendix 1, the friction loss per 100 ft for the given column and tube size is 2.9 ft. The friction loss for the entire 185-ft column length is:

   $$\frac{(2.9 \text{ ft loss}) (185 \text{ ft})}{100 \text{ ft}} = 5.4 \text{ ft}$$

   Lab head can now be determined:

   $$\text{Lab Head,} \atop \text{ft} = 175 \text{ ft} + 5.4 \text{ ft}$$

   $$= 180.4 \text{ ft}$$

PRACTICE PROBLEM 8.5—Cont'd

5. To determine the field head used in the whp calculation, the discharge head must be known:

   (4.6 psi) (2.31 ft/psi) = 10.6 ft

   The field head can now be calculated:

   180 ft + 10.6 ft = 190.6 ft

   And then the water horsepower can be calculated:

   $$whp = \frac{(190.6 \text{ ft})\,(850 \text{ gpm})}{3960}$$

   $$= 40.9 \text{ whp}$$

6. First determine discharge head, in ft, so that field head may be calculated:

   (4.8 psi) (2.31 ft/psi) = 11 ft

   The field head is therefore:

   205 ft + 11 ft = 216 ft

   And then the water horsepower can be calculated:

   $$whp = \frac{(216 \text{ ft})\,(1000 \text{ gpm})}{3960}$$

   $$= 54.5 \text{ whp}$$

7. $$\text{Bowl bhp} = \frac{(\text{Bowl Head, ft})\,(\text{Capac., gpm})}{(3960)\,\dfrac{(\text{Bowl effic.})}{100}}$$

   $$= \frac{(178 \text{ ft})\,(750 \text{ gpm})}{(3960)\,\dfrac{(82)}{100}}$$

   $$= 41 \text{ bowl bhp}$$

8. Bowl bhp = $\dfrac{\text{(Bowl Head, ft) (Capac., gpm)}}{(3960) \dfrac{\text{(Bowl effic.)}}{100}}$

   $= \dfrac{(192 \text{ ft}) (850 \text{ gpm})}{(3960) \dfrac{(78)}{100}}$

   = 52.8 bowl bhp

9. Field bhp = Bowl bhp + Shaft Loss, hp

   Before field bhp can be calculated, the shaft loss must be determined. The table indicates that the shaft loss is 1.14 hp per 100 ft of shaft. The total shaft loss is therefore:

   $\dfrac{(1.14 \text{ hp loss})}{100 \text{ ft}} (170 \text{ ft}) = 1.9 \text{ hp loss}$

   Field bhp can now be calculated:

   Field bhp = Bowl bhp + Shaft Loss, hp

   = 54.2 bhp + 1.9 hp

   = 56.1 bhp

10. The shaft loss, in hp, must be determined before field bhp can be calculated:

    $\dfrac{(0.67 \text{ hp loss})}{100 \text{ ft}} (172 \text{ ft}) = 1.2 \text{ hp loss}$

    Field horsepower can be determined:

    Field bhp = Bowl bhp + Shaft Loss, hp

    = 57.6 bhp + 1.2 hp

    = 58.8 bhp

11. Input or motor hp is calculated as:

    mhp = $\dfrac{\text{Total bhp}}{\dfrac{\text{Motor Effic.}}{100}}$

    $= \dfrac{59.6 \text{ bhp} + 0.5 \text{ hp}}{\dfrac{90}{100}}$

    = 66.8 mhp

PRACTICE PROBLEM 8.5—Cont'd

12. Since the performance curve is based on performance <u>per stage</u>, the bowl head per stage must be calculated:

$$\frac{\text{Bowl Head}}{\text{per Stage}} = \frac{\text{Bowl Head, ft}}{\text{No. of Stages}}$$

$$= \frac{192 \text{ ft}}{4 \text{ stages}}$$

$$= 48 \text{ ft/stage}$$

Now read the performance curve provided in Appendix 2. Find 1100 gpm along the bottom scale and 48 ft along the left scale. The lines from these numbers intersect at about <u>82.4%</u>.

13. The bowl head per stage is:

$$\frac{\text{Bowl Head}}{\text{per Stage}} = \frac{\text{Bowl Head, ft}}{\text{No. of Stages}}$$

$$= \frac{189 \text{ ft}}{3 \text{ stages}}$$

$$= 63 \text{ ft/stage}$$

Now read the performance curve provided in Appendix 2. The intersection of 725 gpm and 63 ft of head is about <u>75%</u>.

14. $$\frac{\text{Field Effic.,}}{\%} = \frac{\text{whp}}{\text{Total Input}} \times 100$$

$$= \frac{44.6 \text{ hp}}{55.1 \text{ bhp}} \times 100$$

$$= 81\%$$

PRACTICE PROBLEM 8.5—Cont'd

15. Before overall efficiency can be calculated, the input hp must be calculated:

$$\text{Input hp} = \frac{\text{Total bhp}}{\dfrac{\text{Motor Effic.}}{100}}$$

$$= \frac{53.7 \text{ bhp}}{\dfrac{88}{100}}$$

$$= 61 \text{ hp Input}$$

Overall efficiency can now be determined:

$$\text{Overall Efficiency, \%} = \frac{42.1 \text{ whp}}{61 \text{ Input hp}} \times 100$$

$$= 69\%$$

PRACTICE PROBLEM 8.6

1. $\text{Volume, gal} = (300 \text{ ft})(120 \text{ ft})(12 \text{ ft})(7.48 \text{ gal/cu ft})$

   $= 3,231,360 \text{ gal}$

2. $\text{Volume, gal} = (350 \text{ ft})(105 \text{ ft})\underbrace{(28 \text{ ft})(0.4)}_{\substack{\text{Estim. Aver.}\\\text{Depth}}}(7.48 \text{ gal/cu ft})$

   $= 3,078,768 \text{ gal}$

3. $\text{Volume, ac-ft} = \dfrac{(190 \text{ ft})(75 \text{ ft})(10 \text{ ft})}{43,560 \text{ cu ft/ac-ft}}$

   $= 3.3 \text{ ac-ft}$

4. $\text{Volume, ac-ft} = \dfrac{(345 \text{ ft})(185 \text{ ft})(23 \text{ ft})(0.4)}{43,560 \text{ cu ft/ac-ft}}$

   $= 13.5 \text{ ac-ft}$

PRACTICE PROBLEM 8.7

1. $\quad$ Copper Sulfate, lbs $= \dfrac{(0.5 \text{ mg}/L \text{ Cu}) (25 \text{ MG}) (8.34 \text{ lbs/gal})}{\dfrac{25}{100}}$

$\qquad\qquad\qquad = 417$ lbs copper sulfate

2. First convert ac-ft volume to gallon volume:

$$(58 \text{ ac-ft}) (43{,}560 \tfrac{\text{cu ft}}{\text{ac-ft}}) (7.48 \text{ gal/cu ft}) = 18{,}898{,}070 \text{ gal}$$

Then calculate lbs copper sulfate required:

$$\dfrac{(0.5 \text{ mg}/L)(18.9 \text{ MG})(8.34 \text{ lbs/gal})}{0.25} = 315.3 \text{ lbs copper sulfate}$$

3. $\quad \dfrac{(0.9 \text{ lbs CuSO}_4) (35 \text{ ac-ft})}{1 \text{ ac-ft}} = 32.4 \text{ lbs copper sulfate}$

4. First calculate ac-ft volume of the pond:

$$\text{Volume, ac-ft} = \dfrac{(240 \text{ ft}) (90 \text{ ft}) (12 \text{ ft})}{43{,}560 \text{ cu ft/ac-ft}}$$

$$= 6.0 \text{ ac-ft}$$

Then calculate lbs copper sulfate required:

$$\dfrac{(0.9 \text{ lbs CuSO}_4) (6 \text{ ac-ft})}{1 \text{ ac-ft}} = 5.9 \text{ lbs copper sulfate}$$

5. First calculate acres area of the storage reservoir:

$$\text{Volume, ac-ft} = \dfrac{(400 \text{ ft}) (120 \text{ ft})}{43{,}560 \text{ sq ft/ac}}$$

$$= 1.1 \text{ ac}$$

Then calculate lbs copper sulfate required:

$$\dfrac{(5.4 \text{ lbs CuSO}_4) (1.1 \text{ ac})}{1 \text{ ac}} = 5.9 \text{ lbs copper sulfate}$$

CHAPTER 8 ACHIEVEMENT TEST

1. Drawdown = Pumping Water Level − Static Water Level

$$= 129.4 \text{ ft} - 92.6 \text{ ft}$$

$$= 36.8 \text{ ft}$$

CHAPTER 8 ACHIEVEMENT TEST—Cont'd

2. $\dfrac{707 \text{ gallons}}{5 \text{ minutes}}$ = 141 gpm

   (141 gpm) (60 min/hr) = 8,460 gph

3. $\dfrac{(0.785)\,(1\text{ ft})\,(1\text{ ft})\,(10\text{ ft})\,(7.48\text{ gal/cu ft})\,(9\text{ round trips})}{5 \text{ min}}$ = 106 gpm

4. First calculate the depth of water in the sounding line:

   (3.7 psi) (2.31 ft/psi) = 8.5 ft water depth
   in sounding line

   $\dfrac{\text{Pumping}}{\text{Water Level, ft}}$ = $\dfrac{\text{Length of}}{\text{Sounding Line, ft}}$ − $\dfrac{\text{Water Depth in}}{\text{Sounding Line, ft}}$

   = 169 ft − 8.5 ft

   = 160.5 ft

   Drawdown can now be calculated:

   Drawdown, ft = $\dfrac{\text{Pumping Water}}{\text{Level, ft}}$ − $\dfrac{\text{Static Water}}{\text{Level, ft}}$

   = 160.5 ft − 139 ft

   = 21.5 ft drawdown

5. $\dfrac{590 \text{ gpm}}{26 \text{ ft drawdown}}$ = 22.7 gpm/ft

6. First calculate the volume of the water-filled casing:

   (0.785) (0.5 ft) (0.5 ft) (140 ft) (7.48 gal/cu ft) = 206 gal

   Then calculate the lbs chlorine required:

   (50 mg/L) (0.000206 MG) (8.34 lbs/gal) = 0.09 lbs Chlorine

7. $\dfrac{760 \text{ gal}}{5 \text{ min}}$ = 152 gpm

   Now convert gpm to gpd:

   $\dfrac{(152 \text{ gal})}{\text{min}}\,\dfrac{(60 \text{ min})}{\text{hr}}\,\dfrac{(8 \text{ hrs})}{\text{day}}$ = 72,960 gal/day

CHAPTER 8 ACHIEVEMENT TEST—Cont'd

8. $(x \text{ mg/}L)$ (0.00059 MG) $(8.34 \text{ lbs/gal}) = 0.49 \text{ lbs}$

$$x = \frac{0.49}{(0.00059)(8.34)}$$

$$= 99.6 \text{ mg/}L$$

9.   First calculate the volume of the water-filled casing:

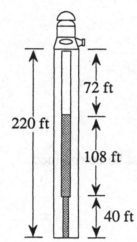

72 ft

220 ft

108 ft

40 ft

(0.785) (1 ft) (1 ft) (108 ft) $(7.48 \text{ gal/cu ft}) = 634 \text{ gal}$

(0.785) (0.67 ft) (0.67 ft) (40 ft) $(7.48 \text{ gal/cu ft}) = \underline{105 \text{ gal}}$

739 gal total

Then calculate the lbs chlorine required:

$(100 \text{ mg/}L)$ (0.000739 MG) $(8.34 \text{ lbs/gal}) = 0.6 \text{ lbs Chlorine}$

10.   First calculate lbs sodium hypochlorite required:

$$\frac{0.1 \text{ lbs chlorine}}{\dfrac{5.25}{100}} = 1.9 \text{ lbs}$$

Next, convert to gal sodium hypochlorite:

$$\frac{1.9 \text{ lbs}}{8.34 \text{ lbs/gal}} = 0.23 \text{ gal}$$

Then convert gal to fl oz:

$$(0.23 \text{ gal}) \left(128 \frac{\text{fl oz}}{\text{gal}}\right) = 29 \text{ fl oz}$$

11. (4.3 psi) $(2.31 \text{ ft/psi}) = 9.9 \text{ ft}$

12. (3.9 psi) $(2.31 \text{ ft/psi}) = 9 \text{ ft}$

```
   98 ft Static Water Level
   27 ft Drawdown
+   9 ft Discharge Head
  134 ft Field Head
```

CHAPTER 8 ACHIEVEMENT TEST—Cont'd

13. Before lab head can be determined, column friction loss must be determined, using the table provided.

    Column friction loss = 4.5 ft/100 ft of shaft

    The friction loss for the entire length of shaft would be:

    $$\frac{(4.5 \text{ ft loss})}{100 \text{ ft}} (190 \text{ ft}) = 8.6 \text{ ft friction loss}$$

    Lab head can now be determined:

    Lab Head, ft = Field head + Column friction loss

              = 180 ft + 8.6 ft

              = 188.6 ft

14. Before water horsepower can be determined, field head must be determined:

    Field head = 187 ft + (4.8 psi) (2.31 ft/psi)

              = 187 ft + 9.9 ft

              = 196.9 ft

    Water horsepower can be calculated:

    $$whp = \frac{(\text{Field Head, ft}) (\text{Capac., gpm})}{3960}$$

    $$whp = \frac{(196.9 \text{ ft}) (900 \text{ gpm})}{3960}$$

    $$= 44.8 \text{ whp}$$

15. $$\text{Bowl bhp} = \frac{(\text{Bowl Head, ft}) (\text{Capac., gpm})}{(3960) \frac{(\text{Bowl effic.})}{100}}$$

    $$= \frac{(178 \text{ ft}) (825 \text{ gpm})}{(3960) (0.80)}$$

    $$= 46.4 \text{ bowl bhp}$$

CHAPTER 8 ACHIEVEMENT TEST—Cont'd

16. Before field bhp can be calculated, the shaft loss (in hp) must be determined using the table.

   Shaft loss = 1.14 hp/100-ft shaft

   The total shaft loss would be:

   $$\frac{(1.14 \text{ hp loss}) (174 \text{ ft})}{100 \text{ ft}} = 2.0 \text{ hp loss}$$

   Field bhp can now be calculated:

   Field bhp = Bowl bhp + Shaft Loss, hp

   $$= 56.4 \text{ bhp} + 2 \text{ hp}$$

   $$= 58.4 \text{ bhp}$$

17.   Input hp $= \dfrac{\text{Total bhp}}{\dfrac{\text{Motor Effic.}}{100}}$

   $$= \frac{49.6 \text{ bhp} + 0.6 \text{ hp}}{0.90}$$

   $$= \frac{50.2}{0.90}$$

   $$= 55.8 \text{ hp Input}$$

18. The intersection of 64 ft of head (192 ft/3 stages) and 675 gpm is a point indicating about 72.5% efficiency.

19. Field Effic., % $= \dfrac{\text{whp}}{\text{Total Input}} \times 100$

   $$= \frac{44.8 \text{ hp}}{56.2 \text{ bhp}} \times 100$$

   $$= 79.6\% \text{ Efficiency}$$

20. Before overall efficiency can be calculated, the input hp must be calculated:

   Input hp $= \dfrac{\text{Total bhp}}{\dfrac{\text{Motor Effic.}}{100}}$

   $$= \frac{53.9 \text{ bhp}}{0.90}$$

   $$= 59.9 \text{ hp Input}$$

   Overall efficiency can now be determined:

   Overall Efficiency, % $= \dfrac{43.5 \text{ whp}}{59.9 \text{ Input hp}} \times 100$

   $$= 72.6 \% \text{ Overall efficiency}$$

CHAPTER 8 ACHIEVEMENT TEST—Cont'd

21. First convert 51 ac-ft to cu ft:

$$(51 \text{ ac-ft}) \left(43,560 \frac{\text{cu ft}}{\text{ac-ft}}\right) = 2,221,560 \text{ cu ft}$$

Then to gallons:
$$(2,221,560 \text{ cu ft}) (7.48 \text{ gal/cu ft}) = 16,617,268 \text{ gal}$$

Then calculate lbs copper sulfate required:

$$\frac{(\text{mg}/L)(\text{MG Vol.})(8.34 \text{ lbs/gal})}{\dfrac{\% \text{ Avail. Cu}}{100}} = \text{lbs copper sulfate}$$

$$\frac{(0.5 \text{ mg}/L)(16.6 \text{ MG})(8.34 \text{ lbs/gal})}{0.25} = \begin{array}{l} 277 \text{ lbs} \\ \text{copper sulfate} \end{array}$$

22. First calculate acres area of the reservoir:

$$\begin{array}{l} \text{Area,} \\ \text{acres} \end{array} = \frac{(420 \text{ ft}) (130 \text{ ft})}{43,560 \text{ sq ft/ac}}$$

$$= 1.3 \text{ ac}$$

Then determine the copper sulfate required:

$$\frac{(5.4 \text{ lbs CuSO}_4) (1.3 \text{ ac})}{1 \text{ ac}} = \begin{array}{l} 7 \text{ lbs lbs} \\ \text{copper sulfate} \end{array}$$

Chapter 9

PRACTICE PROBLEMS 9.1

1. Vol, gal = (4 ft)(3 ft)(3 ft)(7.48 gal/cu ft)
    $$= 269 \text{ gal}$$

2. Vol, gal = (45 ft)(15 ft)(9 ft)(7.48 gal/cu ft)
    $$= 45,441 \text{ gal}$$

3. Vol, gal = (35 ft)(15 ft)(8 ft)(7.48 gal/cu ft)
    $$= 31,416 \text{ gal}$$

4. $\dfrac{40 \text{ in.}}{12 \text{ in./ft}}$ = 3.3 ft

   Vol, gal = (4 ft)(4 ft)(3.3 ft)(7.48 gal/cu ft)

   = 395 gal

5. $\dfrac{10 \text{ in.}}{12 \text{ in./ft}}$ = 0.8 ft

   Vol, gal = (35 ft)(20 ft)(8.8 ft)(7.48 gal/cu ft)

   = 46,077 gal

PRACTICE PROBLEMS 9.2

1. Since detention time is desired in minutes, the flow rate must be expressed as gpm:

   $\dfrac{3,540,000 \text{ gpd}}{1440 \text{ min/day}}$ = 2458 gpm

   Now calculate the detention time:

   $\text{Detention Time, min} = \dfrac{\text{Volume of Tank, gal}}{\text{Flow Rate, gpm}}$

   $= \dfrac{(50 \text{ ft})(20 \text{ ft})(8 \text{ ft})(7.48 \text{ gal/cu ft})}{2458 \text{ gpm}}$

   = 24.3 min

2. First convert the flow rate to gpm:

   $\dfrac{2,600,000 \text{ gpd}}{1440 \text{ min/day}}$ = 1806 gpm

   Then calculate the detention time:

   $\text{Detention Time, min} = \dfrac{\text{Volume of Tank, gal}}{\text{Flow Rate, gpm}}$

   $= \dfrac{(45 \text{ ft})(15 \text{ ft})(9 \text{ ft})(7.48 \text{ gal/cu ft})}{1806 \text{ gpm}}$

   = 25.2 min

PRACTICE PROBLEM 9.2—Cont'd

3. First, convert the flow rate to gps:

$$\frac{8,000,000 \text{ gpd}}{(1440 \text{ min/day})(60 \text{ sec/min})} = 92.6 \text{ gps}$$

Then calculate the detention time:

$$\text{Detention Time, sec} = \frac{(5 \text{ ft})(4 \text{ ft})(4.5 \text{ ft})(7.48 \text{ gal/cu ft})}{92.6 \text{ gps}}$$

$$= 7.3 \text{ sec}$$

4. First, convert the flow rate to gpm:

$$\frac{2,100,000 \text{ gpd}}{1440 \text{ min/day}} = 1458 \text{ gpm}$$

Then calculate the detention time: (8 in. + 12 in. ft = 0.7 ft)

$$\text{Detention Time, sec} = \frac{(40 \text{ ft})(15 \text{ ft})(8.7 \text{ ft})(7.48 \text{ gal/cu ft})}{1458 \text{ gpm}}$$

$$= 26.8 \text{ min}$$

5. Convert the flow rate to gps:

$$\frac{3,150,000 \text{ gpd}}{(1440 \text{ min/day})(60 \text{ sec/min})} = 36.5 \text{ gps}$$

And calculate the detention time: (40 in. = 3.3 ft)

$$\text{Detention Time, sec} = \frac{(3 \text{ ft})(3 \text{ ft})(3.3 \text{ ft})(7.48 \text{ gal/cu ft})}{36.5 \text{ gps}}$$

$$= 6 \text{ sec}$$

PRACTICE PROBLEM 9.3

1. (mg/L Alum)(MGD Flow)(8.34 lbs/gal) = lbs/day Alum

   (9 mg/L)(3.41 MGD)(8.34 lbs/gal) = 256 lbs/day

2. (mg/L Polymer)(MGD Flow)(8.34 lbs/gal) = lbs/day Polymer

   (13 mg/L)(1.726 MGD)(8.34 lbs/gal) = 187 lbs/day

3. (mg/L Alum)(MGD Flow)(8.34 lbs/gal) = lbs/day Alum

   (11 mg/L)(2.82 MGD)(8.34 lbs/gal) = 259 lbs/day

4. (mg/L Alum)(MGD Flow)(8.34 lbs/gal) = lbs/day Alum

   (8 mg/L)(0.96 MGD)(8.34 lbs/gal) = 64 lbs/day

5. (mg/*L* Polymer)(MGD Flow)(8.34 lbs/gal) = lbs/day Polymer

(14 mg/*L*)(4.05 MGD)(8.34 lbs/gal) = 473 lbs/day

PRACTICE PROBLEM 9.4

1. First calculate the lbs/day of dry alum required:

(mg/*L* Alum)(MGD Flow)(8.34 lbs/gal) = lbs/day Dry Alum

(8 mg/*L*)(1.82 MGD)(8.34 lbs/gal) = 121 lbs/day Dry Alum

Then calculate gpd solution required. For each 5.36 lbs of dry alum required, 1 gallon solution is required. Therefore the total gallons solution required is:

$$\frac{121 \text{ lbs/day dry alum}}{5.36 \text{ lbs alum/gal solution}} = 22.6 \text{ gpd Alum Solution}$$

2. Desired Dose, lbs/day = Actual Dose, lbs/day

(mg/*L*)(MGD Flow)(8.34) = (mg/*L*)(MGD Sol'n)(8.34)
Chem. Treated lbs/gal   Sol'n   Flow   lbs/gal

(13 mg/*L*)(3.12 MGD)(8.34 lbs/gal) = (550,000 mg/*L*)(*x* MGD)(8.34 lbs/gal)

$$\frac{(13)(3.12)(\cancel{8.34})}{(550,000)(\cancel{8.34})} = x \text{ MGD}$$

0.0000737 MGD = *x*

MGD flow rate can then be expressed as gpd:

73.7 gpd   = *x*
Alum Solution

3. First determine the lbs/day dry alum required:

(mg/*L* Alum)(MGD Flow)(8.34 lbs/gal) = lbs/day Dry Alum

(11 mg/*L*)(4.02 MGD)(8.34 lbs/gal) = 369 lbs/day Dry Alum

Then calculate the gpd alum solution required:

$$\frac{369 \text{ lbs/day Dry Alum}}{5.45 \text{ lbs Alum/gal solution}} = 68 \text{ gpd solution}$$

PRACTICE PROBLEM 9.4—Cont'd

4. Desired Dose, lbs/day = Actual Dose, lbs/day

$$\underset{\text{Chem. Treated}}{(\text{mg}/L)}(\text{MGD Flow})\underset{\text{lbs/gal}}{(8.34)} = \underset{\text{Sol'n}}{(\text{mg}/L)}(\text{MGD Sol'n})\underset{\text{lbs/gal}}{(8.34)}$$

$$(10 \text{ mg}/L)(0.94 \text{ MGD})(8.34 \text{ lbs/gal}) = (600,000 \text{ mg}/L)(x \text{ MGD})(8.34 \text{ lbs/gal})$$

$$\frac{(10)(0.94)(\cancel{8.34})}{(600,000)(\cancel{8.34})} = x \text{ MGD}$$

$$0.0000156 \text{ MGD} = x$$

MGD flow rate can then be expressed as gpd:

$$\underset{\text{Alum Solution}}{15.6 \text{ gpd}} = x$$

5. First calculate the mg/L concentration of alum in the solution:

$$\frac{642 \text{ mg Alum}}{1 \text{ m}L \text{ Solution}} \times \frac{1000}{1000} = \frac{642,000 \text{ mg Alum}}{1,000 \text{ m}L} = \frac{642,000 \text{ mg Alum}}{1 L}$$

$$\text{or} = 642,000 \text{ mg}/L$$

Then calculate the gpd setting required for the solution chemical feeder:

Desired Dose, lbs/day = Actual Dose, lbs/day

$$(9 \text{ mg}/L)(1.675 \text{ MGD})(8.34 \text{ lbs/gal}) = (642,000 \text{ mg}/L)(x \text{ MGD})(8.34 \text{ lbs/gal})$$

$$\frac{(9)(1.675)(\cancel{8.34})}{(642,000)(\cancel{8.34})} = x \text{ MGD}$$

$$0.0000234 \text{ MGD} = x$$

$$\text{or } 23.4 \text{ gpd} = x$$

PRACTICE PROBLEM 9.5

1. $$\frac{(\text{gpd})(3785 \text{ m}L/\text{gal})}{1440 \text{ min/day}} = \text{m}L/\text{min}$$

$$\frac{(38 \text{ gpd})(3785 \text{ m}L/\text{gal})}{1440 \text{ min/day}} = 100 \text{ m}L/\text{min}$$

2. $$\frac{(\text{gpd})(3785 \text{ m}L/\text{gal})}{1440 \text{ min/day}} = \text{m}L/\text{min}$$

$$\frac{(24.7 \text{ gpd})(3785 \text{ m}L/\text{gal})}{1440 \text{ min/day}} = 65 \text{ m}L/\text{min}$$

3. First calculate the feeder setting in gpd:

$$\frac{(mg/L)(MGD\ Flow)(8.34)}{Chem.\quad Treated\quad lbs/gal} = \frac{(mg/L)(MGD\ Sol'n)(8.34)}{Sol'n\quad Flow\quad lbs/gal}$$

$$(12\ mg/L)(2.92\ MGD)(8.34\ lbs/gal) = (600,000\ mg/L)(x\ MGD)(8.34\ lbs/gal)$$

$$\frac{(12)(2.92)(8.34)}{(600,000)(8.34)} = x\ MGD$$

$$0.0000584\ MGD = x$$
$$or\ 58.4\ gpd = x$$

Then convert gpd setting to mL/min setting:

$$\frac{(58.4\ gpd)(3785\ mL/gal)}{1440\ min/day} = 154\ mL/min$$

4. First calculate the feeder setting in gpd:

$$\frac{(mg/L)(MGD\ Flow)(8.34)}{Chem.\quad Treated\quad lbs/gal} = \frac{(mg/L)(MGD\ Sol'n)(8.34)}{Sol'n\quad Flow\quad lbs/gal}$$

$$(8\ mg/L)(2.67\ MGD)(8.34\ lbs/gal) = (580,000\ mg/L)(x\ MGD)(8.34\ lbs/gal)$$

$$\frac{(8)(2.67)(8.34)}{(580,000)(8.34)} = x\ MGD$$

$$0.0000368\ MGD = x$$
$$or\ 36.8\ gpd = x$$

Then convert gpd setting to mL/min setting:

$$\frac{(36.8\ gpd)(3785\ mL/gal)}{1440\ min/day} = 96.7\ mL/min$$

5. First, calculate the desired gpd setting:

$$\frac{(11\ mg/L)(3.61\ MGD)(8.34\ lbs/gal)}{5.42\ lbs\ Alum/gal\ solution} = 61.1\ gpd$$

Then convert gpd to mL/min:

$$\frac{(61.1\ gpd)(3785\ mL/gal)}{1440\ min/day} = 161\ mL/min$$

PRACTICE PROBLEMS 9.6

1. First determine the lbs dry polymer used in the solution:

$$1g = 0.0022 \text{ lbs}$$

$$130 \text{ g} = (0.0022 \text{ lbs})(130)$$

$$= 0.3 \text{ lbs dry polymer}$$

Then calculate the percent strength of the solution:

$$\frac{0.3 \text{ lbs}}{(15 \text{ gal})(8.34 \text{ lbs/gal}) + 0.3 \text{ lbs}} \times 100 = 0.2\%$$

2. First convert ounces dry polymer used to lbs dry polymer:
   (1 lb = 16 oz, dry measure)

$$\frac{20 \text{ oz}}{16 \text{ oz/lb}} = 1.25 \text{ lbs dry polymer}$$

Then calculate the percent strength of the solution:

$$\frac{1.25 \text{ lbs}}{(20 \text{ gal})(8.34 \text{ lbs/gal}) + 1.25 \text{ lbs}} \times 100 = 0.7\%$$

3. $$\frac{1.9 \text{ lbs}}{(x \text{ gal})(8.34 \text{ lbs/gal}) + 1.9 \text{ lbs}} \times 100 = 0.9$$

$$\frac{(1.9)(100)}{(x)(8.34) + 1.9} = 0.9$$

$$\frac{190}{8.34x + 1.9} = 0.9$$

$$190 = (0.9)(8.34x + 1.9)$$

$$190 = 7.5x + 1.71$$

$$188.29 = 7.5x$$

$$25 \text{ gal} = x$$

4. $$\frac{(10)}{100}(x \text{ lbs liq. polymer}) = \frac{(0.6)}{100}(170 \text{ lbs polymer sol'n})$$

$$0.1 x = (0.006)(170)$$

$$x = \frac{(0.006)(170)}{0.1}$$

$$x = 10.2 \text{ lbs liq. polymer}$$

5. First calculate the density of the liquid polymer solution:

$$(1.3)(8.34 \text{ lbs/gal}) = 10.8 \text{ lbs/gal}$$

Then calculate the gallons liquid polymer required:

$$\frac{(9)}{100}(x \text{ gal liq. polymer})(10.8 \text{ lbs/gal}) = \frac{(0.3)}{100}(40 \text{ gal polymer sol'n})(8.34 \text{ lbs/gal})$$

$$(0.09)(x)(10.8) = (0.003)(40)(8.34)$$

$$x = \frac{(0.003)(40)(8.34)}{(0.09)(10.8)}$$

$$x = 1 \text{ gal liq. polymer}$$

6.
$$\frac{(11)}{100}(x \text{ gal})(10.1 \text{ lbs/gal}) = \frac{(0.7)}{100}(75 \text{ gal})(8.34 \text{ lbs/gal})$$

$$(0.11)(x)(10.1) = (0.007)(75)(8.34)$$

$$x = \frac{(0.007)(75)(8.34)}{(0.11)(10.1)}$$

$$x = 3.9 \text{ gallons liquid polymer}$$

PRACTICE PROBLEMS 9.7

1.
$$\frac{\dfrac{(10)}{100}(30 \text{ lbs}) + \dfrac{(0.5)}{100}(70 \text{ lbs})}{30 \text{ lbs} + 70 \text{ lbs}} \times 100$$

$$= \frac{3 \text{ lbs} + 0.4 \text{ lbs}}{100 \text{ lbs}} \times 100$$

$$= 3.4\% \text{ Strength}$$

2.
$$\frac{\dfrac{(15)}{100}(5 \text{ gal})(10.9 \text{ lbs/gal}) + \dfrac{(0.25)}{100}(45 \text{ gal})(8.38 \text{ lbs/gal})}{(5 \text{ gal})(10.9 \text{ lbs/gal}) + (45 \text{ gal})(8.38 \text{ lbs/gal})} \times 100$$

$$= \frac{8.2 \text{ lbs} + 0.9 \text{ lbs}}{54.5 \text{ lbs} + 377.1 \text{ lbs}} \times 100$$

$$= \frac{9.1 \text{ lbs}}{431.6 \text{ lbs}} \times 100$$

$$= 2.1\% \text{ Strength}$$

PRACTICE PROBLEMS 9.7—Cont'd

3. $$\cfrac{\cfrac{(12)}{100}(10\ gal)(10.5\ lbs/gal) + \cfrac{(0.75)}{100}(40\ gal)(8.42\ lbs/gal)}{(10\ gal)(10.5\ lbs/gal) + (40\ gal)(8.42\ lbs/gal)} \times 100$$

$$= \frac{12.6\ lbs + 2.5\ lbs}{105\ lbs + 336.8\ lbs} \times 100$$

$$= \frac{15.1\ lbs}{441.8\ lbs} \times 100$$

$$= 3.4\%\ Strength$$

4. Use the dilution rectangle:

7   ⟶   2

    4

2   ⟶   3
        ___
       5 parts

7% sol'n: $\dfrac{(2)(400\ lbs)}{5} = 160\ lbs\ of\ 7\%\ Sol'n$

2% sol'n: $\dfrac{(3)(400\ lbs)}{5} = 240\ lbs\ of\ 2\%\ Sol'n$

5. 1   ⟶   0.7

       0.7

   0   ⟶   0.3
           ___
          1.0 parts

1% sol'n: $\dfrac{(0.7)(75\ lbs)}{1} = 52.5\ lbs\ of\ 1\%\ Sol'n$

Water: $\dfrac{(0.3)(75\ lbs)}{1} = 22.5\ lbs\ of\ Water$

6. 8   ⟶   2.5

       3

   0.5   ⟶   5
             ___
            7.5 parts

8% Sol'n: $\dfrac{(2.5)(200\ lbs)}{7.5} = 66.7\ lbs\ of\ 8\%\ Sol'n$

0.5% Sol'n: $\dfrac{(5)(200\ lbs)}{7.5} = 133.3\ lbs\ of\ 0.5\%\ Sol'n$

PRACTICE PROBLEMS 9.8

1. First calculate lbs/min feed rate:

$$\frac{2.1\ lbs}{30\ min} = 0.07\ lbs/min$$

Then convert lbs/min to lbs/day feed rate:

$$(0.07\ lbs/min)(1440\ min/day) = 100.8\ lbs/day$$

2. $\dfrac{40 \text{ oz}}{16 \text{ oz/lb}}$ = 2.5 lbs

    First calculate lbs/min feed rate:

    $\dfrac{2.5 \text{ lbs}}{45 \text{ min}}$ = 0.06 lbs/min

    Then convert lbs/min to lbs/day feed rate:

    (0.06 lbs/min)(1440 min/day) = 86.4 lbs/day

3. $\dfrac{12 \text{ oz}}{16 \text{ oz/lb}}$ = 0.75 lbs bucket

    Determine lbs of chemical used:

    $\quad$ 2.2 $\,$ lbs bucket + chemical
    $-$ 0.75 lbs bucket
    $\quad$ 1.45 lbs chemical

    Then calculate the lbs/min feed rate:

    $\dfrac{1.45 \text{ lbs chemical}}{30 \text{ min}}$ = 0.048 lbs/min

    And convert lbs/min feed rate to lbs/day:

    (0.048 lbs/min)(1440 min/day) = 69 lbs/day

4. Calculate the lbs chemical used:

    $\quad$ 2.7 lbs bucket + chemical
    $-$ 0.5 lbs bucket
    $\quad$ 2.2 lbs chemical

    Next, calculate the lbs/min feed rate:

    $\dfrac{2.2 \text{ lbs chemical}}{30 \text{ min}}$ = 0.073 lbs/min

    Then convert lbs/min to lbs/day:

    (0.073 lbs/min)(1440 min/day) = 105 lbs/day

PRACTICE PROBLEMS 9.8—Cont'd

5. $(x \text{ mg}/L)(1.806 \text{ MGD})(8.34 \text{ lbs/gal}) = 35 \text{ lbs Polymer}$

$$x = \frac{35}{(1.806)(8.34)}$$

$$= 2.3 \text{ mg}/L$$

PRACTICE PROBLEMS 9.9

1. The solution feed rate is 75 gpd. With this information and the concentration (1.5% = 15,000 mg/L), the feed rate in lbs/day can be calculated:

    $(15,000 \text{ mg}/L)(0.000075 \text{ MGD})(8.34 \text{ lbs/gal}) = 9.4 \text{ lbs/day}$

2. The mL/min feed rate is:

    $$\frac{620 \text{ ml}}{5 \text{ min}} = 124 \text{ mL/min}$$

    Convert mL/min solution feed rate to gpd feed rate:

    $$(124 \underline{\text{ mL}}) \ \underline{(1 \text{ gal})} \ (1440 \underline{\text{ min}}) = 47.2 \text{ gpd}$$
    $$\text{min} \quad 3785 \text{ mL} \qquad \text{day}$$

    The polymer feed rate in lbs/day can now be determined:

    $(13,000 \text{ mg}/L)(0.0000472 \text{ MGD})(8.34 \text{ lbs/gal})(1.2 \text{ sp. grav.}) = 6.1 \text{ lbs/day}$

3. The mL/min solution feed rate is:

    $$\frac{710 \text{ ml}}{5 \text{ min}} = 142 \text{ mL/min}$$

    Convert mL/min solution feed rate to gpd feed rate:

    $$(142 \underline{\text{ mL}}) \ \underline{(1 \text{ gal})} \ (1440 \underline{\text{ min}}) = 54 \text{ gpd}$$
    $$\text{min} \quad 3785 \text{ mL} \qquad \text{day}$$

    The lbs/day feed rate can now be determined:

    $(11,000 \text{ mg}/L)(0.000054 \text{ MGD})(8.34 \text{ lbs/gal}) = 5 \text{ lbs/day}$

4. First calculate the mL/min solution feed rate:

$$\frac{930 \text{ m}L}{5 \text{ min}} = 186 \text{ m}L/\text{min}$$

Convert mL/min solution feed rate to gpd feed rate:

$$\frac{(186 \text{ m}L)}{\text{min}} \frac{(1 \text{ gal})}{3785 \text{ m}L} \frac{(1440 \text{ min})}{\text{day}} = 70.8 \text{ gpd}$$

The lbs/day polymer feed can now be calculated:

(13,000 mg/L)(0.0000708 MGD)(8.34 lbs/gal) = 7.7 lbs/day

5. The mL/min solution feed flow is:

$$\frac{1680 \text{ ml}}{10 \text{ min}} = 168 \text{ m}L/\text{min}$$

And the mL/min flow is then converted to gpd:

$$\frac{(168 \text{ m}L)}{\text{min}} \frac{(1 \text{ gal})}{3785 \text{ m}L} \frac{(1440 \text{ min})}{\text{day}} = 63.9 \text{ gpd}$$

And the lbs/day polymer feed rate is therefore:

(18,000 mg/L)(0.0000639 MGD)(8.34 lbs/gal)(1.09 sp. grav.) = 10.5 lbs/day

PRACTICE PROBLEMS 9.10

1. $\dfrac{4 \text{ in.}}{12 \text{ in./ft}} = 0.3 \text{ ft}$

The gpm pumping rate is:

$$\frac{(0.785)(3 \text{ ft})(3 \text{ ft})(0.3 \text{ ft})(7.48 \text{ gal/cu ft})}{5 \text{ min}} = 3.2 \text{ gpm}$$

2. $\dfrac{3 \text{ in.}}{12 \text{ in./ft}} = 0.25 \text{ ft}$

The gpm pumping rate is:

$$\frac{(0.785)(3 \text{ ft})(3 \text{ ft})(0.25 \text{ ft})(7.48 \text{ gal/cu ft})}{10 \text{ min}} = 1.3 \text{ gpm}$$

PRACTICE PROBLEMS 9.10—Cont'd

3. $\dfrac{2 \text{ in.}}{12 \text{ in./ft}} = 0.17 \text{ ft}$

   The gpm pumping rate is:

   $\dfrac{(0.785)(3 \text{ ft})(3 \text{ ft})(0.17 \text{ ft})(7.48 \text{ gal/cu ft})}{10 \text{ min}} = 0.9 \text{ gpm}$

   The gpd pumping rate is therefore:

   $(0.9 \text{ gpm})(1440 \text{ min/day}) = 1296 \text{ gpd}$

4. First calculate the gpm solution flow: (1.5 in. ÷ 12 in./ft = 0.13 ft)

   $\dfrac{(0.785)(3 \text{ ft})(3 \text{ ft})(0.13 \text{ ft})(7.48 \text{ gal/cu ft})}{20 \text{ min}} = 0.3 \text{ gpm}$

   Next, determine the gpd pumping rate:

   $(0.3 \text{ gpm})(1440 \text{ min/day}) = 432 \text{ gpd}$

   And then calculate the polymer feed rate, lbs/day:

   $(12,000 \text{ mg}/L)(0.000432 \text{ MGD})(8.34 \text{ lbs/gal}) = 43.2 \text{ lbs/day}$

5. The gpm solution flow is: (1 in. ÷ 12 in./ft = 0.08 ft)

   $\dfrac{(0.785)(3 \text{ ft})(3 \text{ ft})(0.08 \text{ ft})(7.48 \text{ gal/cu ft})}{30 \text{ min}} = 0.14 \text{ gpm}$

   And the gpd solution flow is:

   $(0.14 \text{ gpm})(1440 \text{ min/day}) = 202 \text{ gpd}$

   Therefore, the polymer feed rate, lbs/day:

   $(12,500 \text{ mg}/L)(0.000202 \text{ MGD})(8.34 \text{ lbs/gal}) = 21 \text{ lbs/day}$

PRACTICE PROBLEMS 9.11

1. $\dfrac{530 \text{ lbs}}{7 \text{ days}} = 76 \text{ lbs/day average}$

2. $\dfrac{2100 \text{ lbs}}{95 \text{ lbs/day}} = 22.1 \text{ days}$

3. $\dfrac{921 \text{ lbs}}{60 \text{ lbs/day}} = 15.4 \text{ days}$

4. First calculate the gallons supply:

   (0.785)(4 ft)(4 ft)(3.2 ft)(7.48 gal/cu ft) = 301 gal

   Then calculate days' supply:

   $$\frac{301 \text{ gal}}{92 \text{ gpd}} = 3.3 \text{ days}$$

5. (2.5 mg/L)(1.6 MGD)(8.34 lbs/gal) = 33.4 lbs/day

   The polymer used during a 30-day period would therefore be:

   (33.4 lbs/day)(30 days) = 1002 lbs

CHAPTER 9 ACHIEVEMENT TEST

1. Since detention time is desired in seconds, the flow should be converted to gps:

   $$\frac{5,700,000 \text{ gpd}}{(1440 \text{ min/day})(60 \text{ sec/min})} = 66 \text{ gps}$$

   The detention time can now be calculated:

   $$\frac{\text{Detention Time}}{\text{Sec}} = \frac{(4 \text{ ft})(3 \text{ ft})(3 \text{ ft})(7.48 \text{ gal/cu ft})}{66 \text{ gps}}$$

   $$= 4 \text{ sec}$$

2. Vol, gal = (40 ft)(15 ft)(8 ft)(7.48 gal/cu ft)

   = 35,904 gal

3. (7 mg/L)(3.62 MGD)(8.34 lbs/gal) = 211 lbs/day

4. (9 mg/L)(2.95 MGD)(8.34 lbs/gal) = (550,000 mg/L)(x MGD)(8.34 lbs/gal)

   $$\frac{(9)(2.95)(\cancel{8.34})}{(550,000)(\cancel{8.34})} = x$$

   0.0000482 MGD = x

   or 48.2 gpd = x

5. (4 ft)(4 ft)(3 ft)(7.48 gal/cu ft) = 359 gal

6. $$\frac{(40 \text{ gpd})(3785 \text{ m}L/\text{gal})}{1440 \text{ min/day}} = 105 \text{ m}L/\text{min}$$

CHAPTER 9 ACHIEVEMENT TEST—Cont'd

7. Since detention time is desired in minutes, express the flow rate in gpm:

$$\frac{2,480,000 \text{ gpd}}{1440 \text{ min/day}} = 1772 \text{ gpm}$$

$$\text{Detention Time, min} = \frac{(45 \text{ ft}) (20 \text{ ft}) (9.75 \text{ ft}) (7.48 \text{ gal/cu ft})}{1722 \text{ gpm}}$$

$$= 38 \text{ min}$$

8. First determine the gpd feeder setting desired:

$$(7 \text{ mg}/L)(1.96 \text{ MGD})(8.34 \text{ lbs/gal}) = (600,000 \text{ mg}/L)(x \text{ MGD})(10.2 \text{ lbs/gal})$$

$$\frac{(7)(1.96)(8.34)}{(600,000)(10.2)} = x \text{ MGD}$$

$$0.0000187 \text{ MGD} = x$$

$$\text{or } 18.7 \text{ gpd} = x$$

Then convert gpd feeder setting to m*L*/min feeder setting:

$$\frac{(18.7 \text{ gpd})(3785 \text{ m}L/\text{gal})}{1440 \text{ min/day}} = 49.2 \text{ m}L/\text{min}$$

9. $$\frac{(85 \text{ gpd})(3785 \text{ m}L/\text{gal})}{1440 \text{ min/day}} = 223 \text{ m}L/\text{min}$$

10. $$(8 \text{ mg}/L)(0.94 \text{ MGD})(8.34 \text{ lbs/gal}) = 62.7 \text{ lbs/day}$$

11. $$\frac{2.5 \text{ lbs}}{(x \text{ gal})(8.34 \text{ lbs/gal}) + 2.5 \text{ lbs}} \times 100 = 1.3$$

$$\frac{250}{8.34x + 2.5} = 1.3$$

$$\frac{250}{1.3} = 8.34x + 2.5$$

$$192.3 = 8.34x + 2.5$$

$$189.8 = 8.34x$$

$$\frac{189.8}{8.34} = x$$

$$22.8 \text{ gal} = x$$

12. $$\frac{\dfrac{(15)(20\ lbs)}{100} + \dfrac{(0.6)(130\ lbs)}{100}}{20\ lbs + 130\ lbs} \times 100$$

$$= \frac{3\ lbs + 0.78\ lbs}{150\ lbs} \times 100$$

$$= \frac{3.78\ lbs}{150\ lbs} \times 100$$

$$= 2.5\%$$

13. $$\frac{3.5\ lbs\ chemical}{30\ min} \ \frac{\times\ 2}{\times\ 2} = \frac{7\ lbs}{60\ min} \ \text{or 7 lbs/hr}$$

Then $(7\ \frac{lbs}{hr})(24\ \frac{hrs}{day}) = 168\ lbs/day$

14.   3.9 lbs bucket + chemical
   − 1.5 lbs bucket
    2.4 lbs chemical

The feed rate is therefore:

$$\frac{2.4\ lbs}{30\ min} \ \frac{\times\ 2}{\times\ 2} = \frac{4.8\ lbs}{60\ min} = 4.8\ lbs/hr$$

Then $(4.8\ \frac{lbs}{hr})(24\ \frac{hrs}{day}) = 115.2\ lbs/day$

15.   1 gram = 0.0022 lbs

  180 grams = (0.0022 lbs)(180)

        = 0.4 lbs

The percent strength of the solution can now be determined:

$$\frac{0.4\ lbs}{(20\ gal)(8.34\ lbs/gal) + 0.4\ lbs} \times 100$$

$$= \frac{0.4}{166.8 + 0.4} \times 100$$

$$= \frac{0.4}{167.2} \times 100$$

$$= 0.2\%$$

CHAPTER 9 ACHIEVEMENT TEST—Cont'd

16. The mL/min solution feed rate is:

$$\frac{750 \text{ m}L}{5 \text{ min}} = 150 \text{ m}L/\text{min}$$

The solution feed rate must then be converted to gpd:

$$\frac{(150 \text{ m}L/\text{min})(1440 \text{ min}/\text{day})}{3785 \text{ m}L/\text{gal}} = 57 \text{ gpd}$$

The dry polymer feed rate can now be determined:

$$(15,000 \text{ mg}/L)(0.000057 \text{ MGD})(8.34 \text{ lbs/gal}) = 7.1 \text{ lbs/day}$$

17. First calculate the lbs/day dry alum required:

$$(12 \text{ mg}/L)(4.1 \text{ MGD})(8.34 \text{ lbs/gal}) = 410 \text{ lbs/day}$$

Then calculate the gpd alum solution required:

$$\frac{410 \text{ lbs/day}}{5.36 \text{ lbs alum/gal sol'n}} = 76.5 \text{ gpd}$$

18. Use the dilution rectangle:

10 ⟶ 0.7   10% sol'n: $\frac{(0.7)(200 \text{ lbs})}{10}$ = 14 lbs of 10% Sol'n

0.7

0 ⟶ 9.3   Water: $\frac{(9.3)(200 \text{ lbs})}{10}$ = 186 lbs of Water

10 parts

19. 55 ⟶ 1   55% sol'n: $\frac{(1)(150 \text{ lbs})}{55}$ = 2.7 lbs of 55% Sol'n

1

0 ⟶ 54   Water: $\frac{(54)(150 \text{ lbs})}{55}$ = 147.3 lbs of Water

55 parts

20. $\frac{2 \text{ in.}}{12 \text{ in./ft}} = 0.17 \text{ ft}$

The gpm pumping rate is:

$$\frac{(0.785)(3 \text{ ft})(3 \text{ ft})(0.17 \text{ ft})(7.48 \text{ gal/cu ft})}{10 \text{ min}} = \frac{9 \text{ gal}}{10 \text{ min}}$$

$$= 0.9 \text{ gpm}$$

21. $\dfrac{(10)(x \text{ gal})(9.8 \text{ lbs/gal})}{100} = \dfrac{(0.5)(75 \text{ gal})(8.34 \text{ lbs/gal})}{100}$

$$x = \dfrac{(0.005)(75)(8.34)}{(0.1)(9.8)}$$

$$= 3.2 \text{ gal}$$

22. The mL/min solution feed rate is:

$$\dfrac{680 \text{ mL}}{5 \text{ min}} = 136 \text{ mL/min}$$

This can then be converted to gpd:

$$\dfrac{(136 \text{ mg/L/min})(1440 \text{ min/day})}{3785 \text{ mL/gal}} = 51.7 \text{ gpd}$$

The lbs/day polymer dosage rate can be calculated:

$$(9000 \text{ mg/L})(0.0000517 \text{ MGD})(8.34 \text{ lbs/gal}) = 3.9 \text{ lbs/day}$$

23. $(5 \text{ mg/L})(3.4 \text{ MGD})(8.34 \text{ lbs/gal}) = 142 \text{ lbs/day}$

The polymer required for 30 days would be:

$$\left(142 \dfrac{\text{lbs}}{\text{day}}\right)(30 \text{ days}) = 4260 \text{ lbs}$$

24. $\dfrac{530 \text{ lbs}}{70 \text{ lbs/day}} = 7.6 \text{ days}$

Chapter 10

PRACTICE PROBLEMS 10.1

1. Vol., gal = (length, ft)(width, ft)(depth, ft)(7.48 gal/cu ft)

    = (60 ft)(25 ft)(12 ft)(7.48 gal/cu ft)

    = 134,640 gal

2. Vol., gal = $(0.785)(D^2)$(depth, ft)(7.48 gal/cu ft)

    = (0.785)(70 ft)(70 ft)(10 ft)(7.48 gal/cu ft)

    = 287,718 gal

PRACTICE PROBLEMS 10.1—Cont'd

3. Vol., gal = (75 ft)(25 ft)(12 ft)(7.48 gal/cu ft)

    = 168,300 gal

4. 60,000 gal = (50 ft)(20 ft)(x ft)(7.48 gal/cu ft)

    $$\frac{60,000}{(50)(20)(7.48)} = x$$

    8 ft = x

5. $\dfrac{3\ in.}{12\ in./ft} = 0.25\ ft$

    Vol., gal = (0.785)(70 ft)(70 ft)(10.25 ft)(7.48 gal/cu ft)

    = 294,911 gal

PRACTICE PROBLEMS 10.2

1. First convert gpd flow rate to gph flow rate:

    $$\frac{2,120,000\ gpd}{24\ hrs/day} = 88,333\ gph$$

    Then calculate detention time, in hours:

    $$\text{Detention Time, hrs} = \frac{(80\ ft)(30\ ft)(12\ ft)(7.48\ gal/cu\ ft)}{88,333\ gph}$$

    = 2.4 hrs

2. First convert gpd flow rate to gph flow rate:

    $$\frac{2,815,000\ gpd}{24\ hrs/day} = 117,292\ gph$$

    Then calculate detention time, in hours:

    $$\text{Detention Time, hrs} = \frac{(0.785)(70\ ft)(70\ ft)(11\ ft)(7.48\ gal/cu\ ft)}{117,292\ gph}$$

    = 2.7 hrs

3. Convert gpd flow rate to gph flow rate:

$$\frac{1,470,000 \text{ gpd}}{24 \text{ hrs/day}} = 61,250 \text{ gph}$$

Then calculate detention time:

$$\text{Detention Time, hrs} = \frac{(50 \text{ ft})(15 \text{ ft})(12 \text{ ft})(7.48 \text{ gal/cu ft})}{61,250 \text{ gph}}$$

$$= 1.1 \text{ hrs}$$

4. Use the detention time equation, fill in the given data, then solve for the unknown variable:

$$\text{Detention Time, hrs} = \frac{\text{Volume of Tank, gal}}{\text{Flow Rate, gph}}$$

$$2.5 \text{ hrs} = \frac{(0.785)(50 \text{ ft})(50 \text{ ft})(10 \text{ ft})(7.48 \text{ gal/cu ft})}{x \text{ gph}}$$

$$x = \frac{(0.785)(50)(50)(10)(7.48)}{2.5}$$

$$x = 58,718 \text{ gph}$$

Then convert gph flow to gpd flow:

$$(58,718 \text{ gph})(24 \text{ hrs/day}) = 1,409,232 \text{ gpd}$$

$$\text{or} = 1.4 \text{ MGD}$$

5. First convert gpd flow rate to gph flow rate:

$$\frac{1,570,000 \text{ gpd}}{24 \text{ hrs/day}} = 65,417 \text{ gph}$$

Then calculate detention time, in hours:

$$\text{Detention Time, hrs} = \frac{(60 \text{ ft})(20 \text{ ft})(10 \text{ ft})(7.48 \text{ gal/cu ft})}{65,417 \text{ gph}}$$

$$= 1.4 \text{ hrs}$$

PRACTICE PROBLEMS 10.3

1. Surface Overflow Rate, gpm/sq ft $= \dfrac{490 \text{ gpm}}{(50 \text{ ft})(20 \text{ ft})}$

$= 0.49 \text{ gpm/sq ft}$

2. Surface Overflow Rate, gpm/sq ft $= \dfrac{1550 \text{ gpm}}{(0.785)(60 \text{ ft})(60 \text{ ft})}$

$= 0.55 \text{ gpm/sq ft}$

3. First convert gpd flow rate to gpm flow rate:

$\dfrac{530,000 \text{ gpd}}{1440 \text{ min/day}} = 368 \text{ gpm}$

Then calculate surface overflow rate in gpm/sq ft:

Surface Overflow Rate, gpm/sq ft $= \dfrac{368 \text{ gpm}}{(45 \text{ ft})(20 \text{ ft})}$

$= 0.4 \text{ gpm/sq ft}$

4. First calculate gpm flow rate using the surface overflow rate equation:

$0.5 \text{ gpm/sq ft} = \dfrac{x \text{ gpm}}{(75 \text{ ft})(25 \text{ ft})}$

$(0.5)(75)(25) = x \text{ gpm}$

$937.5 \text{ gpm} = x$

Now convert gpm flow to gpd flow:

$(937.5 \text{ gpm})(1440 \text{ min/day}) = 1,350,000 \text{ gpd}$

5. First convert gpd flow to gpm flow:

$\dfrac{1,670,000 \text{ gpd}}{1440 \text{ min/day}} = 1160 \text{ gpm}$

Then calculate surface overflow rate:

Surface Overflow Rate, gpm/sq ft $= \dfrac{1160 \text{ gpm}}{(0.785)(50 \text{ ft})(50 \text{ ft})}$

$= 0.59 \text{ gpm/sq ft}$

PRACTICE PROBLEMS 10.4

1. Convert gpd flow rate to cfm flow rate:

$$\frac{1,440,000 \text{ gpd}}{(1440 \text{ min/day}) (7.48 \text{ gal/cu ft})} = 134 \text{ cfm}$$

Then calculate ft/min velocity:

$$Q_{cfm} = A V_{fpm}$$

$$134 \text{ cfm} = (20 \text{ ft}) (12 \text{ ft}) (x \text{ fpm})$$

$$\frac{134}{(20)(12)} = x$$

$$0.6 \text{ fpm} = x$$

2. Convert gpd flow rate to cfm flow rate:

$$\frac{1,600,000 \text{ gpd}}{(1440 \text{ min/day}) (7.48 \text{ gal/cu ft})} = 149 \text{ cfm}$$

Then calculate ft/min velocity:

$$Q_{cfm} = A V_{fpm}$$

$$149 \text{ cfm} = (25 \text{ ft}) (10 \text{ ft}) (x \text{ fpm})$$

$$\frac{149}{(25)(10)} = x$$

$$0.6 \text{ fpm} = x$$

3. Convert gpd flow rate to cfm flow rate:

$$\frac{2,240,000 \text{ gpd}}{(1440 \text{ min/day}) (7.48 \text{ gal/cu ft})} = 208 \text{ cfm}$$

Then calculate ft/min velocity:

$$Q_{cfm} = A V_{fpm}$$

$$208 \text{ cfm} = (30 \text{ ft}) (13 \text{ ft}) (x \text{ fpm})$$

$$\frac{208}{(30)(13)} = x$$

$$0.5 \text{ fpm} = x$$

PRACTICE PROBLEMS 10.4—Cont'd

4. First convert flow rate to proper terms:

$$\frac{2{,}950{,}000 \text{ gpd}}{(1440 \text{ min/day}) (7.48 \text{ gal/cu ft})} = 274 \text{ cfm}$$

Then calculate ft/min velocity:

$$Q_{cfm} = A \, V_{fpm}$$

$$274 \text{ cfm} = (30 \text{ ft}) (12 \text{ ft}) (x \text{ fpm})$$

$$\frac{274}{(30) \, (12)} = x$$

$$0.8 \text{ fpm} = x$$

5. Convert flow rate to cfm:

$$\frac{890{,}000 \text{ gpd}}{(1440 \text{ min/day}) (7.48 \text{ gal/cu ft})} = 83 \text{ cfm}$$

Then calculate ft/min velocity:

$$Q_{cfm} = A \, V_{fpm}$$

$$83 \text{ cfm} = (20 \text{ ft}) (10 \text{ ft}) (x \text{ fpm})$$

$$\frac{83}{(20) \, (10)} = x$$

$$0.4 \text{ fpm} = x$$

PRACTICE PROBLEMS 10.5

1. First convert gpd flow rate to gpm flow rate:

$$\frac{2{,}420{,}000 \text{ gpd}}{1440 \text{ min/day}} = 1681 \text{ gpm}$$

Then calculate weir loading rate:

$$\begin{aligned} \text{Weir Loading Rate} &= \frac{\text{Flow, gpm}}{\text{Weir Length, ft}} \\ &= \frac{1681 \text{ gpm}}{(3.14)(60 \text{ ft})} \\ &= 8.9 \text{ gpm/ft} \end{aligned}$$

2. Convert gpd flow rate to gpm flow rate:

$$\frac{1,940,000 \text{ gpd}}{1440 \text{ min/day}} = 1347 \text{ gpm}$$

Then calculate weir loading rate:

$$\text{Weir Loading Rate} = \frac{\text{Flow, gpm}}{\text{Weir Length, ft}}$$

$$= \frac{1347 \text{ gpm}}{170 \text{ ft}}$$

$$= 7.9 \text{ gpm/ft}$$

3. First convert gpd flow rate to gpm flow rate:

$$\frac{1,260,000 \text{ gpd}}{1440 \text{ min/day}} = 875 \text{ gpm}$$

Then calculate weir loading rate:

$$\text{Weir Loading Rate} = \frac{\text{Flow, gpm}}{\text{Weir Length, ft}}$$

$$= \frac{875 \text{ gpm}}{110 \text{ ft}}$$

$$= 8.0 \text{ gpm/ft}$$

4. Convert gpd flow rate to gpm flow rate:

$$\frac{3,100,000 \text{ gpd}}{1440 \text{ min/day}} = 2153 \text{ gpm}$$

Then calculate weir loading rate:

$$\text{Weir Loading Rate} = \frac{\text{Flow, gpm}}{\text{Weir Length, ft}}$$

$$= \frac{2153 \text{ gpm}}{(3.14)(75 \text{ ft})}$$

$$= 9.1 \text{ gpm/ft}$$

PRACTICE PROBLEMS 10.5—Cont'd

5. First convert flow rate to gpm:

$$\frac{1,800,000 \text{ gpd}}{1440 \text{ min/day}} = 1250 \text{ gpm}$$

Then calculate weir loading rate:

$$\text{Weir Loading Rate} = \frac{\text{Flow, gpm}}{\text{Weir Length, ft}}$$

$$= \frac{1250 \text{ gpm}}{150 \text{ ft}}$$

$$= 8.3 \text{ gpm/ft}$$

PRACTICE PROBLEMS 10.6

1. $\%$ Settled Sludge $= \dfrac{\text{Settled Sludge, m}L}{\text{Total Sample, m}L} \times 100$

$$= \frac{21 \text{ m}L}{100 \text{ m}L} \times 100$$

$$= 21\%$$

2. $\%$ Settled Sludge $= \dfrac{\text{Settled Sludge, m}L}{\text{Total Sample, m}L} \times 100$

$$= \frac{23 \text{ m}L}{100 \text{ m}L} \times 100$$

$$= 23\%$$

3. $\%$ Settled Sludge $= \dfrac{\text{Settled Sludge, m}L}{\text{Total Sample, m}L} \times 100$

$$= \frac{17 \text{ m}L}{100 \text{ m}L} \times 100$$

$$= 17\%$$

4. % Settled
   Sludge
   $= \dfrac{\text{Settled Sludge, m}L}{\text{Total Sample, m}L} \times 100$

   $= \dfrac{18 \text{ m}L}{100 \text{ m}L} \times 100$

   $= 18\%$

PRACTICE PROBLEMS 10.7

1. First calculate the mg/L alkalinity that will react with 48 mg/L alum:

   $$\dfrac{0.45 \text{ mg/}L \text{ Alk.}}{1 \text{ mg/}L \text{ Alum}} = \dfrac{x \text{ mg/}L \text{ Alk.}}{48 \text{ mg/}L \text{ Alum}}$$

   $$(0.45)(48) = x$$

   $$21.6 \text{ mg/}L = x$$

   Then calculate the total alkalinity required:

   $\begin{aligned} \text{Total Alk.} \\ \text{Req'd, mg/}L \end{aligned} = \begin{aligned} \text{Alk. to React with} \\ \text{Alum, mg/}L \end{aligned} + \begin{aligned} \text{"Residual" Alk.,} \\ \text{mg/}L \end{aligned}$

   $= 21.6 \text{ mg/}L + 30 \text{ mg/}L$

   $= 51.6 \text{ mg/}L$

2. First calculate the mg/L alkalinity that will react with 50 mg/L alum:

   $$\dfrac{0.45 \text{ mg/}L \text{ Alk.}}{1 \text{ mg/}L \text{ Alum}} = \dfrac{x \text{ mg/}L \text{ Alk.}}{50 \text{ mg/}L \text{ Alum}}$$

   $$(0.45)(50) = x$$

   $$22.5 \text{ mg/}L = x$$

   Then calculate the total alkalinity required:

   $\begin{aligned} \text{Total Alk.} \\ \text{Req'd, mg/}L \end{aligned} = \begin{aligned} \text{Alk. to React with} \\ \text{Alum, mg/}L \end{aligned} + \begin{aligned} \text{"Residual" Alk.,} \\ \text{mg/}L \end{aligned}$

   $= 22.5 \text{ mg/}L + 30 \text{ mg/}L$

   $= 52.5 \text{ mg/}L$

PRACTICE PROBLEMS 10.7—Cont'd

3. Alk. to be Added to the Water, mg/L = Total Alk. Req'd, mg/L − Alk. Present in the Water, mg/L

    = 45 mg/L − 28 mg/L

    = 17 mg/L Alk. to be Added

4. Alk. to be Added to the Water, mg/L = Total Alk. Req'd, mg/L − Alk. Present in the Water, mg/L

    = 42 mg/L − 26 mg/L

    = 16 mg/L Alk. to be Added

5. $\dfrac{0.45 \text{ mg/}L \text{ Alk.}}{0.35 \text{ mg/}L \text{ Lime}} = \dfrac{17 \text{ mg/}L \text{ Alk.}}{x \text{ mg/}L \text{ Lime}}$

    $0.45\,x = (17)\,(0.35)$

    $x = \dfrac{(17)\,(0.35)}{0.45}$

    $x = 13.2 \text{ mg/}L \text{ Lime}$

6. $\dfrac{0.45 \text{ mg/}L \text{ Alk.}}{0.35 \text{ mg/}L \text{ Lime}} = \dfrac{21 \text{ mg/}L \text{ Alk.}}{x \text{ mg/}L \text{ Lime}}$

    $0.45\,x = (21)\,(0.35)$

    $x = \dfrac{(21)\,(0.35)}{0.45}$

    $x = 16.3 \text{ mg/}L \text{ Lime}$

7. To determine the total alkalinity required, you must first calculate the alkalinity that will react with 52 mg/*L* alum:

$$\frac{0.45 \text{ mg/}L \text{ Alk.}}{1 \text{ mg/}L \text{ Alum}} = \frac{x \text{ mg/}L \text{ Alk.}}{52 \text{ mg/}L \text{ Alum}}$$

$$(0.45)(52) = x$$

$$23.4 \text{ mg/}L = x$$
$$\text{Alk.}$$

Now the total alkalinity requirement can be calculated:

$$\begin{array}{ccc} \text{Total Alk.} \\ \text{Req'd, mg/}L \end{array} = \begin{array}{c} \text{Alk. to React} \\ \text{with Alum, mg/}L \end{array} + \begin{array}{c} \text{"Residual"} \\ \text{Alk., mg/}L \end{array}$$

$$= 23.4 \text{ mg/}L + 30 \text{ mg/}L$$

$$= \begin{array}{c} 53.4 \text{ mg/}L \text{ Total} \\ \text{Alk. Required} \end{array}$$

Next, calculate how much alkalinity must be <u>added</u> to the water:

$$\begin{array}{c} \text{Alk. to be Added} \\ \text{to the Water, mg/}L \end{array} = \begin{array}{c} \text{Total Alk.} \\ \text{Req'd, mg/}L \end{array} - \begin{array}{c} \text{Alk. Present} \\ \text{in the Water, mg/}L \end{array}$$

$$= 53.4 \text{ mg/}L - 36 \text{ mg/}L$$

$$= \begin{array}{c} 17.4 \text{ mg/}L \text{ Alk.} \\ \text{to be Added to the Water} \end{array}$$

Then calculate the lime required (in mg/*L*) in order to provide this additional alkalinity:

$$\frac{0.45 \text{ mg/}L \text{ Alk.}}{0.35 \text{ mg/}L \text{ Lime}} = \frac{17.4 \text{ mg/}L \text{ Alk.}}{x \text{ mg/}L \text{ Lime}}$$

$$0.45 \, x = (17.4)(0.35)$$

$$x = \frac{(17.4)(0.35)}{0.45}$$

$$x = 13.5 \text{ mg/}L \text{ Lime Required}$$

PRACTICE PROBLEMS 10.8

1. (13.6 mg/*L*)(2.4 MGD)(8.34 lbs/gal) = 272 lbs/day Lime

2. (12.9 mg/*L*)(2.15 MGD)(8.34 lbs/gal) = 231 lbs/day Lime

3. (15.2 mg/*L*)(0.97 MGD)(8.34 lbs/gal) = 123 lbs/day Lime

4. (14 mg/*L*)(1.1 MGD)(8.34 lbs/gal) = 128 lbs/day Lime

PRACTICE PROBLEMS 10.9

1. $\dfrac{(195 \text{ lbs/day}) (453.6 \text{ g/lb})}{1440 \text{ min/day}}$ = 61.4 g/min Lime

2. $\dfrac{(105 \text{ lbs/day}) (453.6 \text{ g/lb})}{1440 \text{ min/day}}$ = 33.1 g/min Lime

3. First calculate lbs/day lime required:

   (14 mg/*L*)(0.81 MGD)(8.34 lbs/gal) = 94.6 lbs/day Lime

   Then convert lbs/day to g/min lime required:

   $\dfrac{(94.6 \text{ lbs/day}) (453.6 \text{ g/lb})}{1440 \text{ min/day}}$ = 29.8 g/min Lime

4. First calculate lbs/day lime required:

   (12.5 mg/*L*)(2.84 MGD)(8.34 lbs/gal) = 296.1 lbs/day Lime

   Then convert lbs/day to g/min lime required:

   $\dfrac{(296.1 \text{ lbs/day}) (453.6 \text{ g/lb})}{1440 \text{ min/day}}$ = 93.3 g/min Lime

CHAPTER 10 ACHIEVEMENT TEST

1. $\dfrac{\text{Detention}}{\text{Time, hrs}}$ = $\dfrac{\text{Volume of Tank, gal}}{\text{Flow Rate, gph}}$

   Since detention time is desired in hours, the flow rate should be expressed as gph:

   $\dfrac{1{,}270{,}000 \text{ gpd}}{24 \text{ hrs/day}}$ = 52,917 gph

   Then the detention time can be calculated:

   $\dfrac{\text{Detention}}{\text{Time, hrs}}$ = $\dfrac{(55 \text{ ft})(20 \text{ ft})(10 \text{ ft})(7.48 \text{ gal/cu ft})}{52{,}917 \text{ gph}}$

   = 1.6 hrs

2. (65 ft)(25 ft)(12 ft)(7.48 gal/cu ft) = 145,860 gal

3. $Q_{cfm} = A V_{fpm}$

First convert gpd flow rate to cfm:

$$\frac{1,510,000 \text{ gpd}}{(1440 \text{ min/day})(7.48 \text{ gal/cu ft})} = 140 \text{ cfm}$$

Then use the Q = AV equation to determine velocity:

$$Q_{cfm} = A V_{fpm}$$

$$140 \text{ cfm} = (20 \text{ ft})(11 \text{ ft})(x \text{ fpm})$$

$$\frac{140}{(20)(11)} = x$$

$$0.6 \text{ fpm} = x$$

4. $\dfrac{\text{Surface Overflow}}{\text{Rate, gpm/sq ft}} = \dfrac{\text{Flow, gpm}}{\text{Filter Surface Area, sq ft}}$

First convert gpd flow rate to gpm flow rate:

$$\frac{615,000 \text{ gpd}}{(1440 \text{ min/day})} = 427 \text{ gpm}$$

Then calculate surface overflow rate:

$$\frac{\text{Surface Overflow}}{\text{Rate, gpm/sq ft}} = \frac{427 \text{ gpm}}{(45 \text{ ft})(20 \text{ ft})}$$

$$= 0.5 \text{ gpm/sq ft}$$

5. $(0.785)(60 \text{ ft})(60 \text{ ft})(12 \text{ ft})(7.48 \text{ gal/cu ft}) = 253,662 \text{ gal}$

6. Since weir loading rate is expressed as gpm/ft, the flow rate must be converted to gpm:

$$\frac{2,110,000 \text{ gpd}}{1440 \text{ min/day}} = 1465 \text{ gpm}$$

Then calculate weir loading rate:

$$\frac{\text{Weir Loading}}{\text{Rate, gpm/ft}} = \frac{1465 \text{ gpm}}{175 \text{ ft}}$$

$$= 8.4 \text{ gpm/ft}$$

CHAPTER 10 ACHIEVEMENT TEST—Cont'd

7. Detention time is desired in hours; therefore, the flow rate should be expressed in gph:

$$\frac{2,920,000 \text{ gpd}}{24 \text{ hrs/day}} = 121,667 \text{ gph}$$

Now calculate the detention time:

$$\text{Detention Time, hrs} = \frac{\text{Volume of Tank, gal}}{\text{Flow Rate, gph}}$$

$$= \frac{(0.785)(70 \text{ ft})(70 \text{ ft})(12 \text{ ft})(7.48 \text{ gal/cu ft})}{121,667 \text{ gph}}$$

$$= 2.8 \text{ hrs}$$

8. $Q_{cfm} = A V_{fpm}$

Since velocity is desired in fpm, express the gpd flow rate as cfm flow rate:

$$\frac{1,720,000 \text{ gpd}}{(1440 \text{ min/day}) (7.48 \text{ gal/cu ft})} = 160 \text{ cfm}$$

Then use the Q = AV equation to determine velocity:

$$Q_{cfm} = A V_{fpm}$$

$$160 \text{ cfm} = (25 \text{ ft}) (10 \text{ ft}) (x \text{ fpm})$$

$$\frac{160}{(25) (10)} = x$$

$$0.6 \text{ fpm} = x$$

9. $$\text{Surface Overflow Rate, gpm/sq ft} = \frac{1600 \text{ gpm}}{(0.785) (60 \text{ ft}) (60 \text{ ft})}$$

$$= 0.6 \text{ gpm/sq ft}$$

10. Weir loading rate is expressed as gpm/ft. Therefore, the flow rate should be expressed as gpm:

$$\frac{2,510,000 \text{ gpd}}{1440 \text{ min/day}} = 1743 \text{ gpm}$$

Then calculate weir loading rate:

$$\text{Weir Loading Rate, gpm/ft} = \frac{1743 \text{ gpm}}{(3.14) (60 \text{ ft})}$$

$$= 9.3 \text{ gpm/ft}$$

11. First calculate flow rate in gph using the detention time equation:

$$\text{Detention Time, hrs} = \frac{\text{Volume, gal}}{\text{Flow Rate, gph}}$$

$$2 \text{ hrs} = \frac{(0.785)(50 \text{ ft})(50 \text{ ft})(11 \text{ ft})(7.48 \text{ gal/cu ft})}{x \text{ gph}}$$

$$x = \frac{(0.785)(50)(50)(11)(7.48)}{2}$$

$$x = 80,737 \text{ gph}$$

The gph flow rate can now be converted to gpd (and then MGD) flow rate:

$$(80,737 \text{ gph}) (24 \text{ hrs/day}) = 1,937,688 \text{ gpd}$$

$$\text{or} = 1.94 \text{ MGD}$$

12. $Q_{cfm} = A V_{fpm}$

Since velocity must be expressed in fpm, the flow rate should be expressed as cfm:

$$\frac{3,120,000 \text{ gpd}}{(1440 \text{ min/day}) (7.48 \text{ gal/cu ft})} = 290 \text{ cfm}$$

The velocity can now be determined using the Q = AV equation:

$$Q_{cfm} = A V_{fpm}$$

$$290 \text{ cfm} = (30 \text{ ft}) (12 \text{ ft}) (x \text{ fpm})$$

$$\frac{290}{(30)(12)} = x$$

$$0.8 \text{ fpm} = x$$

13. $$\frac{\% \text{ Settled}}{\text{Sludge}} = \frac{\text{Settled Sludge Volume, m}L}{\text{Total Sample Volume, m}L} \times 100$$

$$= \frac{24 \text{ m}L}{100 \text{ m}L} \times 100$$

$$= 24\% \text{ Settled Sludge}$$

CHAPTER 10 ACHIEVEMENT TEST—Cont'd

14. $$\begin{array}{ccc} \text{Total Alk.} \\ \text{Required} \\ \text{mg}/L \end{array} = \begin{array}{c} \text{Alk. that will} \\ \text{React with} \\ \text{the Alum, mg}/L \end{array} + \begin{array}{c} \text{Alk. Required to} \\ \text{Assure Proper} \\ \text{Precipitation of} \\ \text{Alum, mg}/L \end{array}$$

First calculate the alkalinity required to react with the alum dose of 50 mg/L:

$$\frac{1 \text{ mg}/L \text{ Alum}}{0.45 \text{ mg}/L \text{ Alk.}} = \frac{50 \text{ mg}/L \text{ Alum}}{x \text{ mg}/L \text{ Alk.}}$$

$$(0.45)(50) = x$$

$$22.5 \text{ mg}/L = x$$

Then calculate the total alkalinity required:

$$\begin{array}{c} \text{Total Alk.} \\ \text{Req'd, mg}/L \end{array} = 22.5 \text{ mg}/L + 30 \text{ mg}/L$$

$$= 52.5 \text{ mg}/L$$

15. Use the normal surface overflow rate to determine the maximum flow. The flow rate is first determined as gpm (as given in the equation):

$$\begin{array}{c} \text{Surface Overflow} \\ \text{Rate, gpm/sq ft} \end{array} = \frac{\text{Flow rate, gpm}}{\text{Filter Surface Area, sq ft}}$$

$$0.6 \text{ gpm/sq ft} = \frac{x \text{ gpm}}{(75 \text{ ft}) (25 \text{ ft})}$$

$$(0.6) (75) (25) = x$$

$$1125 \text{ gpm} = x$$

Then the flow rate is converted to gpd:

$$(1125 \text{ gpm}) (1440 \text{ min/day}) = 1{,}620{,}000 \text{ gpd}$$

16. $(14.2 \text{ mg}/L) (2.37 \text{ MGD}) (8.34 \text{ lbs/gal}) = 280.7 \text{ lbs/day}$

17. $$\begin{array}{c} \% \text{ Settled} \\ \text{Sludge} \end{array} = \frac{\text{Settled Sludge Volume, m}L}{\text{Total Sample Volume, m}L} \times 100$$

$$= \frac{19 \text{ m}L}{100 \text{ m}L} \times 100$$

$$= 19\% \text{ Settled Sludge}$$

18. Since weir loading rate is expressed as gpm/ft, the MGD flow rate should be expressed as gpm:

$$\frac{3,130,000 \text{ gpd}}{1440 \text{ min/day}} = 2174 \text{ gpm}$$

Then calculate weir loading rate in gpm/ft:

$$\text{Weir Loading Rate, gpm/ft} = \frac{2174 \text{ gpm}}{(3.14)(75 \text{ ft})}$$

$$= 9.2 \text{ gpm/ft}$$

19.
$$\begin{array}{c}\text{Total Alk.} \\ \text{Req'd, mg}/L\end{array} - \begin{array}{c}\text{Alk. Present} \\ \text{in the Water, mg}/L\end{array} = \begin{array}{c}\text{Alk. to be Added} \\ \text{to the Water, mg}/L\end{array}$$

$$40 \text{ mg}/L - 28 \text{ mg}/L = \begin{array}{c}12 \text{ mg}/L \text{ Alk. to be} \\ \text{added to the water}\end{array}$$

20. To determine total alkalinity required, first calculate the alkalinity that will react with 48 mg/L alum:

$$\frac{0.45 \text{ mg}/L \text{ Alk.}}{1 \text{ mg}/L \text{ Alum}} = \frac{x \text{ mg}/L \text{ Alk.}}{48 \text{ mg}/L \text{ Alum}}$$

$$(0.45)(48) = x$$

$$21.6 \text{ mg}/L \text{ Alk} = x$$

The total alkalinity requirement can now be determined:

$$\begin{array}{c}\text{Total Alk.} \\ \text{Req'd, mg}/L\end{array} = \begin{array}{c}\text{Alk. to React} \\ \text{with Alum, mg}/L\end{array} + \begin{array}{c}\text{"Residual"} \\ \text{Alk., mg}/L\end{array}$$

$$= 21.6 \text{ mg}/L + 30 \text{ mg}/L$$

$$= 51.6 \text{ mg}/L \text{ Alk. Required}$$

Now calculate how much alkalinity must be <u>added</u> to the water:

$$\begin{array}{c}\text{Alk. to be Added} \\ \text{to the Water, mg}/L\end{array} = \begin{array}{c}\text{Total Alk.} \\ \text{Req'd, mg}/L\end{array} - \begin{array}{c}\text{Alk. Present} \\ \text{in the Water, mg}/L\end{array}$$

$$= 51.6 \text{ mg}/L - 32 \text{ mg}/L$$

$$= 19.6 \text{ mg}/L \text{ Alk. to be Added}$$

Then calculate the lime dosage required (in mg/L) to provide this additional alkalinity:

$$\frac{0.45 \text{ mg}/L \text{ Alk.}}{0.35 \text{ mg}/L \text{ Lime}} = \frac{19.6 \text{ mg}/L \text{ Alk.}}{x \text{ mg}/L \text{ Lime}}$$

$$x = \frac{(19.6)(0.35)}{0.45}$$

$$x = 15.2 \text{ mg}/L \text{ Lime Required}$$

CHAPTER 10 ACHIEVEMENT TEST—Cont'd

21. $\dfrac{(188 \text{ lbs/day Lime}) (453.6 \text{ g/lb})}{1440 \text{ min/day}} = 59.2 \text{ g/min}$

22. To determine total alkalinity required, first calculate the alkalinity that will react with 50 mg/*L* alum:

$$\frac{0.45 \text{ mg/}L \text{ Alk.}}{1 \text{ mg/}L \text{ Alum}} = \frac{x \text{ mg/}L \text{ Alk.}}{50 \text{ mg/}L \text{ Alum}}$$

$$(0.45)(50) = x$$

$$22.5 \text{ mg/}L = x$$

The total alkalinity requirement can now be determined:

$$\begin{aligned} \text{Total Alk.} \\ \text{Req'd, mg/}L \end{aligned} = 22.5 \text{ mg/}L + 30 \text{ mg/}L$$

$$= 52.5 \text{ mg/}L$$

Next, calculate how much alkalinity must be <u>added</u> to the water:

$$\begin{aligned} \text{Alk. to be Added} \\ \text{to the Water, mg/}L \end{aligned} = \begin{aligned} \text{Total Alk.} \\ \text{Req'd, mg/}L \end{aligned} - \begin{aligned} \text{Alk. Present} \\ \text{in the Water, mg/}L \end{aligned}$$

$$= 52.5 \text{ mg/}L - 34 \text{ mg/}L$$

$$= 18.5 \text{ mg/}L \text{ Alk. to be Added}$$

Now calculate the lime dosage required to provide this level of alkalinity:

$$\frac{0.45 \text{ mg/}L \text{ Alk.}}{0.35 \text{ mg/}L \text{ Lime}} = \frac{18.5 \text{ mg/}L \text{ Alk.}}{x \text{ mg/}L \text{ Lime}}$$

$$x = \frac{(18.5)(0.35)}{0.45}$$

$$x = 14.4 \text{ mg/}L \text{ Lime Required}$$

23. $(15 \text{ mg/}L)(1.3 \text{ MGD})(8.34 \text{ lbs/gal}) = 163 \text{ lbs/day}$

24. $\dfrac{(\text{Lime, lbs/day})(453.6 \text{ g/lb})}{1440 \text{ min/day}} = \text{Lime, g/min}$

The mg/*L* to lbs/day equation can replace the lbs/day factor:

$$\frac{(\text{Lime, mg/}L)(\text{MGD Flow})(8.34 \text{ lbs/gal})(453.6 \text{ g/lb})}{1440 \text{ min/day}}$$

$$= \frac{(13 \text{ mg/}L)(2.94 \text{ MGD})(8.34 \text{ lbs/gal})(453.6 \text{ g/lb})}{1440 \text{ min/day}}$$

$$= 100 \text{ g/min Lime}$$

Chapter 11

PRACTICE PROBLEMS 11.1

1. Flow Rate, gpm = $\dfrac{\text{Total Gallons Produced}}{\text{Filter Run, min}}$

   = $\dfrac{13,500,000 \text{ gal}}{(80 \text{ hrs})(60 \text{ min/hr})}$

   = 2813 gpm

2. $\dfrac{2,770,000 \text{ gpd}}{1440 \text{ min/day}}$ = 1924 gpm

3. Flow Rate, gpm = $\dfrac{\text{Total Gallons Produced}}{\text{Filter Run, min}}$

   3000 gpm = $\dfrac{14,000,000 \text{ gal}}{(x \text{ hrs})(60 \text{ min/hr})}$

   $x = \dfrac{14,000,000 \text{ gal}}{(3000)(60)}$

   $x = 78 \text{ hrs}$

4. (13 in. ÷ 12 in./ft = 1.1 ft)

   Q_{gpm} = (Length)(Width)(Drop Veloc.)(7.48)
   ft   ft   ft/min   gal/cu ft

   = (40 ft)(25 ft)$\dfrac{(1.1 \text{ ft})}{5 \text{ min}}$$\dfrac{(7.48 \text{ gal})}{\text{cu ft}}$

   = 1646 gpm

5. (16 in. ÷ 12 in./ft = 1.3 ft)

   Q_{gpm} = (Length)(Width)(Drop Veloc.)(7.48)
   ft   ft   ft/min   gal/cu ft

   = (30 ft)(25 ft)$\dfrac{(1.3 \text{ ft})}{5 \text{ min}}$$\dfrac{(7.48 \text{ gal})}{\text{cu ft}}$

   = 1459 gpm

6. (20 in. ÷ 12 in./ft = 1.7 ft)

$$Q_{gpm} = \underset{ft}{(Length)} \; \underset{ft}{(Width)} \; \underset{ft/min}{(Drop\;Veloc.)} \; \underset{gal/cu\;ft}{(7.48)}$$

$$= (40\;ft)\;(20\;ft)\;\frac{(1.7\;ft)}{8\;min}\;\frac{(7.48\;gal)}{cu\;ft}$$

$$= 1272\;gpm$$

PRACTICE PROBLEMS 11.2

1. Filtration Rate, $= \dfrac{1930\;gpm}{(25\;ft)\;(20\;ft)}$
    gpm/sq ft

    $= 3.9\;gpm/sq\;ft$

2. First convert gpd flow rate to gpm flow rate:

    $$\frac{3,360,000\;gpd}{1440\;min/day} = 2333\;gpm$$

    Then calculate filtration rate:

    Filtration Rate, $= \dfrac{2333\;gpm}{(35\;ft)\;(20\;ft)}$
    gpm/sq ft

    $= 3.3\;gpm/sq\;ft$

3. First calculate the gpm flow rate through the filter:

    Flow Rate, $= \dfrac{Total\;Gallons\;Produced}{Filter\;Run,\;min}$
    gpm

    $$= \frac{18,700,000\;gal}{(73.4\;hrs)\;(60\;min/hr)}$$

    $= 4246\;gpm$

    Then calculate filtration rate:

    Filtration Rate, $= \dfrac{4246\;gpm}{(40\;ft)\;(25\;ft)}$
    gpm/sq ft

    $4.2\;gpm/sq\;ft$

PRACTICE PROBLEMS 11.2—Cont'd

4. First calculate the gpm flow rate through the filter:

$$\text{Flow Rate,} \atop \text{gpm} = \frac{\text{Total Gallons Produced}}{\text{Filter Run, min}}$$

$$= \frac{14,900,000 \text{ gal}}{(72.6 \text{ hrs}) (60 \text{ min/hr})}$$

$$= 3421 \text{ gpm}$$

Then determine the filtration rate:

$$\text{Filtration Rate,} \atop \text{gpm/sq ft} = \frac{3421 \text{ gpm}}{(35 \text{ ft}) (25 \text{ ft})}$$

$$3.9 \text{ gpm/sq ft}$$

5. First convert gpd flow rate to gpm flow rate:

$$\frac{3,740,000 \text{ gpd}}{1440 \text{ min/day}} = 2597 \text{ gpm}$$

Then calculate filtration rate:

$$\text{Filtration Rate,} \atop \text{gpm/sq ft} = \frac{2597 \text{ gpm}}{(40 \text{ ft}) (25 \text{ ft})}$$

$$= 2.6 \text{ gpm/sq ft}$$

6. First calculate gpm flow rate using the Q = AV equation:
    (25 in. ÷ 12 in./ft = 2.1 ft)

$$Q_{\text{gpm}} = \underset{\text{ft}}{(\text{Length})} \underset{\text{ft}}{(\text{Width})} \underset{\text{ft/min}}{(\text{Drop Veloc.})} \underset{\text{gal/cu ft}}{(7.48)}$$

$$= (40 \text{ ft}) (20 \text{ ft}) \frac{(2.1 \text{ ft})}{5 \text{ min}} \frac{(7.48 \text{ gal})}{\text{cu ft}}$$

$$= 2513 \text{ gpm}$$

Then calculate filtration rate:

$$\text{Filtration Rate,} \atop \text{gpm/sq ft} = \frac{2513 \text{ gpm}}{(40 \text{ ft}) (20 \text{ ft})}$$

$$= 3.1 \text{ gpm/sq ft}$$

7. First calculate gpm flow rate using the Q = AV equation:
   (23 in. ÷ 12 in./ft = 1.9 ft)

$$Q_{gpm} = \underset{ft}{(Length)} \; \underset{ft}{(Width)} \; \underset{ft/min}{(Drop\;Veloc.)} \; \underset{gal/cu\;ft}{(7.48)}$$

$$= (35\;ft)\;(25\;ft)\;\frac{(1.9\;ft)}{7\;min}\;(7.48\;\frac{gal}{cu\;ft})$$

$$= 1777\;gpm$$

Then calculate filtration rate:

$$\underset{gpm/sq\;ft}{Filtration\;Rate,} = \frac{1777\;gpm}{(35\;ft)\;(25\;ft)}$$

$$= 2\;gpm/sq\;ft$$

PRACTICE PROBLEMS 11.3

1. $UFRV = \dfrac{Total\;Gallons\;Filtered}{Filter\;Surface\;Area,\;sq\;ft}$

   $= \dfrac{2,910,000\;gal}{(20\;ft)\;(20\;ft)}$

   $= 7275\;gal/sq\;ft$

2. $UFRV = \dfrac{Total\;Gallons\;Filtered}{Filter\;Surface\;Area,\;sq\;ft}$

   $= \dfrac{4,140,000\;gal}{(30\;ft)\;(20\;ft)}$

   $= 6900\;gal/sq\;ft$

3. $UFRV = \dfrac{Total\;Gallons\;Filtered}{Filter\;Surface\;Area,\;sq\;ft}$

   $= \dfrac{3,040,000\;gal}{(25\;ft)\;(20\;ft)}$

   $= 6080\;gal/sq\;ft$

4. $UFRV = \underset{gpm/sq\;ft}{(Filtration\;Rate,\;)} \; \underset{min}{(Filter\;Run\;Time,\;)}$

   $= (3.3\;gpm/sq\;ft)\;(3260\;min)$

   $= 10,758\;gal/sq\;ft$

PRACTICE PROBLEMS 11.3—Cont'd

5. UFRV = (Filtration Rate,) (Filter Run,) (60 $\frac{min}{hr}$)
    gpm/sq ft   hrs

    = (2.4 gpm/sq ft) (62.6 hrs) (60 $\frac{min}{hr}$)

    = 9014 gal/sq ft

PRACTICE PROBLEMS 11.4

1. Backwash = $\frac{Flow\ Rate,\ gpm}{Filter\ Area,\ sq\ ft}$
    Rate

    = $\frac{3440\ gpm}{375\ sq\ ft}$

    = 9.2 gpm/sq ft

2. Backwash = $\frac{Flow\ Rate,\ gpm}{Filter\ Area,\ sq\ ft}$
    Rate

    = $\frac{3760\ gpm}{(20\ ft)(15\ ft)}$

    = 12.5 gpm/sq ft

3. $\frac{(18\ gpm/sq\ ft)\ (12\ in./ft)}{7.48\ gal/cu\ ft}$ = 28.9 in./min

4. Backwash = $\frac{3700\ gpm}{(30\ ft)(20\ ft)}$
    Rate

    = 6.2 gpm/sq ft

5. First calculate the backwash rate as gpm/sq ft:

    Backwash = $\frac{3150\ gpm}{(20\ ft)(15\ ft)}$
    Rate

    = 10.5 gpm/sq ft

    Then convert gpm/sq ft to in./min rise:

    (10.5 gpm/sq ft) (1.6) = 16.8 in./min rise

PRACTICE PROBLEMS 11.5

1. $\begin{aligned}\text{Backwash} \\ \text{Water Vol.,} \\ \text{gal}\end{aligned} = \begin{aligned}\text{(Backwash)} \\ \text{Flow Rate,} \\ \text{gpm}\end{aligned} \begin{aligned}\text{(Duration of)} \\ \text{Backwash,} \\ \text{min}\end{aligned}$

$$= (6{,}700 \text{ gpm}) (7 \text{ min})$$

$$= 46{,}900 \text{ gal}$$

2. $\begin{aligned}\text{Backwash} \\ \text{Water Vol.,} \\ \text{gal}\end{aligned} = \begin{aligned}\text{(Backwash)} \\ \text{Flow Rate,} \\ \text{gpm}\end{aligned} \begin{aligned}\text{(Duration of)} \\ \text{Backwash,} \\ \text{min}\end{aligned}$

$$= (9{,}050 \text{ gpm}) (8 \text{ min})$$

$$= 72{,}400 \text{ gal}$$

3. $\begin{aligned}\text{Backwash} \\ \text{Water Vol.,} \\ \text{gal}\end{aligned} = \begin{aligned}\text{(Backwash)} \\ \text{Flow Rate,} \\ \text{gpm}\end{aligned} \begin{aligned}\text{(Duration of)} \\ \text{Backwash,} \\ \text{min}\end{aligned}$

$$= (4{,}860 \text{ gpm}) (6 \text{ min})$$

$$= 29{,}160 \text{ gal}$$

4. $\begin{aligned}\text{Backwash} \\ \text{Water Vol.,} \\ \text{gal}\end{aligned} = \begin{aligned}\text{(Backwash)} \\ \text{Flow Rate,} \\ \text{gpm}\end{aligned} \begin{aligned}\text{(Duration of)} \\ \text{Backwash,} \\ \text{min}\end{aligned}$

$$= (6{,}840 \text{ gpm}) (6 \text{ min})$$

$$= 41{,}040 \text{ gal}$$

PRACTICE PROBLEMS 11.6

1. $\text{Vol., gal} = (0.785) (D^2) (\text{depth}) (7.48 \text{ gal/cu ft})$

$58{,}940 \text{ gal} = (0.785) (45 \text{ ft}) (45 \text{ ft}) (x \text{ ft}) (7.48 \text{ gal/cu ft})$

$$\frac{58{,}940}{(0.785) (45) (45) (7.48)} = x$$

$$5.0 \text{ ft} = x$$

PRACTICE PROBLEMS 11.6—Cont'd

2. Vol., gal $= (0.785) (D^2) (depth) (7.48 \text{ gal/cu ft})$

    $61,700 \text{ gal} = (0.785) (50 \text{ ft}) (50 \text{ ft}) (x \text{ ft}) (7.48 \text{ gal/cu ft})$

    $$\frac{61,700}{(0.785) (50) (50) (7.48)} = x$$

    $4.2 \text{ ft} = x$

3. Vol., gal $= (0.785) (D^2) (depth) (7.48 \text{ gal/cu ft})$

    $41,400 \text{ gal} = (0.785) (40 \text{ ft}) (40 \text{ ft}) (x \text{ ft}) (7.48 \text{ gal/cu ft})$

    $$\frac{41,400}{(0.785) (40) (40) (7.48)} = x$$

    $4.4 \text{ ft} = x$

4. First calculate the volume of backwash water required:

    $$\begin{matrix} \text{Backwash} \\ \text{Water Vol.,} \\ \text{gal} \end{matrix} = \begin{matrix} \text{(Backwash)} \\ \text{Flow Rate,} \\ \text{gpm} \end{matrix} \begin{matrix} \text{(Duration of)} \\ \text{Backwash,} \\ \text{min} \end{matrix}$$

    $= (7,200 \text{ gpm}) (8 \text{ min})$

    $= 57,600 \text{ gal}$

    Then calculate the depth required in the backwash water tank:

    Vol., gal $= (0.785) (D^2) (depth) (7.48 \text{ gal/cu ft})$

    $57,600 \text{ gal} = (0.785) (50 \text{ ft}) (50 \text{ ft}) (x \text{ ft}) (7.48 \text{ gal/cu ft})$

    $$\frac{57,600}{(0.785) (50) (50) (7.48)} = x$$

    $3.9 \text{ ft} = x$

5. First calculate the volume of water required for backwashing:

$$\begin{matrix} \text{Backwash} \\ \text{Water Vol.,} \\ \text{gal} \end{matrix} = \begin{matrix} \text{(Backwash)} \\ \text{Flow Rate,} \\ \text{gpm} \end{matrix} \begin{matrix} \text{(Duration of)} \\ \text{Backwash,} \\ \text{min} \end{matrix}$$

$$= (8,950 \text{ gpm}) (7 \text{ min})$$

$$= 62,650 \text{ gal}$$

Then calculate the depth required in the backwash water tank:

$$\text{Vol., gal} = (0.785) (D^2) (\text{depth}) (7.48 \text{ gal/cu ft})$$

$$62,650 \text{ gal} = (0.785) (45 \text{ ft}) (45 \text{ ft}) (x \text{ ft}) (7.48 \text{ gal/cu ft})$$

$$\frac{62,650}{(0.785) (45) (45) (7.48)} = x$$

$$5.3 \text{ ft} = x$$

PRACTICE PROBLEMS 11.7

1. $$\begin{matrix} \text{Backwash Pumping} \\ \text{Rate, gpm} \end{matrix} = \begin{matrix} \text{(Desired Backwash)} \\ \text{Rate, gpm/sq ft} \end{matrix} \begin{matrix} \text{(Filter Area,)} \\ \text{sq ft} \end{matrix}$$

$$= (18 \text{ gpm/sq ft}) (40 \text{ ft}) (20 \text{ ft})$$

$$= 14,400 \text{ gpm}$$

2. $$\begin{matrix} \text{Backwash Pumping} \\ \text{Rate, gpm} \end{matrix} = \begin{matrix} \text{(Desired Backwash)} \\ \text{Rate, gpm/sq ft} \end{matrix} \begin{matrix} \text{(Filter Area,)} \\ \text{sq ft} \end{matrix}$$

$$= (15 \text{ gpm/sq ft}) (35 \text{ ft}) (25 \text{ ft})$$

$$= 13,125 \text{ gpm}$$

3. $$\begin{matrix} \text{Backwash Pumping} \\ \text{Rate, gpm} \end{matrix} = \begin{matrix} \text{(Desired Backwash)} \\ \text{Rate, gpm/sq ft} \end{matrix} \begin{matrix} \text{(Filter Area,)} \\ \text{sq ft} \end{matrix}$$

$$= (17 \text{ gpm/sq ft}) (20 \text{ ft}) (20 \text{ ft})$$

$$= 6,800 \text{ gpm}$$

4. $$\begin{matrix} \text{Backwash Pumping} \\ \text{Rate, gpm} \end{matrix} = \begin{matrix} \text{(Desired Backwash)} \\ \text{Rate, gpm/sq ft} \end{matrix} \begin{matrix} \text{(Filter Area,)} \\ \text{sq ft} \end{matrix}$$

$$= (24 \text{ gpm/sq ft}) (25 \text{ ft}) (20 \text{ ft})$$

$$= 12,000 \text{ gpm}$$

PRACTICE PROBLEMS 11.8

1. Backwash Water, % $= \dfrac{\text{Backwash Water, gal}}{\text{Water Filtered, gal}} \times 100$

    $= \dfrac{73,700 \text{ gal}}{16,840,000 \text{ gal}} \times 100$

    $= 0.44\%$ Backwash Water

2. Backwash Water, % $= \dfrac{\text{Backwash Water, gal}}{\text{Water Filtered, gal}} \times 100$

    $= \dfrac{36,300 \text{ gal}}{5,860,000 \text{ gal}} \times 100$

    $= 0.62\%$ Backwash Water

3. Backwash Water, % $= \dfrac{\text{Backwash Water, gal}}{\text{Water Filtered, gal}} \times 100$

    $= \dfrac{58,200 \text{ gal}}{12,962,000 \text{ gal}} \times 100$

    $= 0.45\%$ Backwash Water

4. Backwash Water, % $= \dfrac{\text{Backwash Water, gal}}{\text{Water Filtered, gal}} \times 100$

    $= \dfrac{51,710 \text{ gal}}{10,905,000 \text{ gal}} \times 100$

    $= 0.47\%$ Backwash Water

PRACTICE PROBLEMS 11.9

1. The mud ball volume is 517 mL – 500 mL = 17 mL

    % Mud Ball Volume $= \dfrac{\text{Mud Ball Vol., m}L}{\text{Total Sample Vol., m}L} \times 100$

    $= \dfrac{17 \text{ m}L}{3475 \text{ m}L} \times 100$

    $= 0.5\%$

2. The mud ball volume is 529 mL – 500 mL = 29 mL.
   The total sample volume is (5)(695 mL) = 3475 mL

$$\%\ \text{Mud Ball Volume} = \frac{\text{Mud Ball Vol., m}L}{\text{Total Sample Vol., m}L} \times 100$$

$$= \frac{29\ \text{m}L}{3475\ \text{m}L} \times 100$$

$$= 0.8\%$$

3. The mud ball volume is 581 mL – 500 mL = 81 mL.
   The total sample volume is (5)(695 mL) = 3475 mL

$$\%\ \text{Mud Ball Volume} = \frac{\text{Mud Ball Vol., m}L}{\text{Total Sample Vol., m}L} \times 100$$

$$= \frac{81\ \text{m}L}{3475\ \text{m}L} \times 100$$

$$= 2.3\%$$

4. The mud ball volume is 552 mL – 500 mL = 52 mL.
   The total sample volume is (5)(695 mL) = 3475 mL

$$\%\ \text{Mud Ball Volume} = \frac{\text{Mud Ball Vol., m}L}{\text{Total Sample Vol., m}L} \times 100$$

$$= \frac{52\ \text{m}L}{3475\ \text{m}L} \times 100$$

$$= 1.5\%$$

CHAPTER 11 ACHIEVEMENT TEST

1. $$\text{Flow Rate, gpm} = \frac{\text{Total Gallons Produced}}{\text{Filter Run, min}}$$

$$= \frac{12,900,000\ \text{gal}}{(75\ \text{hrs})\ (60\ \text{min/hr})}$$

$$= 2867\ \text{gpm}$$

CHAPTER 11 ACHIEVEMENT TEST—Cont'd

2. First convert MGD flow rate to gpm flow rate:
   (3.26 MGD = 3,260,000 gpd)

   $$\frac{3,260,000 \text{ gpd}}{1440 \text{ min/day}} = 2264 \text{ gpm}$$

   Then calculate filtration rate:

   $$\text{Filtration Rate, gpm/sq ft} = \frac{2264 \text{ gpm}}{(35 \text{ ft}) (20 \text{ ft})}$$

   $$= 3.2 \text{ gpm/sq ft}$$

3. $$\text{UFRV} = \frac{\text{Total Gallons Filtered}}{\text{Filter Surface Area, sq ft}}$$

   $$= \frac{2,640,000 \text{ gal}}{(20 \text{ ft}) (20 \text{ ft})}$$

   $$= 6600 \text{ gal/sq ft}$$

4. $$\text{Flow Rate, gpm} = \frac{\text{Total Gallons Produced}}{\text{Filter Run, min}}$$

   $$2800 \text{ gpm} = \frac{14,300,000 \text{ gal}}{(x \text{ hrs}) (60 \text{ min/hr})}$$

   $$x = \frac{14,300,000}{(2800) (60)}$$

   $$x = 85.1 \text{ hrs}$$

5. First calculate cfm flow rate using the Q = AV equation:
   (Water drop is 15 in. ÷ 12 in./ft = 1.25 ft)

   $$Q_{cfm} = (\text{Length, ft}) (\text{Width, ft}) (\text{Drop Veloc., ft/min})$$

   $$= \frac{(35 \text{ ft}) (25 \text{ ft}) (1.25 \text{ ft})}{5 \text{ min}}$$

   $$= 219 \text{ cfm}$$

   Then convert cfm flow rate to gpm flow rate:

   $$(219 \text{ cfm}) (7.48 \text{ gal/cu ft}) = 1638 \text{ gpm}$$

6. $UFRV = \dfrac{3,260,000 \text{ gal}}{(25 \text{ ft}) (25 \text{ ft})}$

   $= 5216 \text{ gal/sq ft}$

7. First calculate the gpm flow rate through the filter:

   $\begin{matrix} \text{Flow Rate,} \\ \text{gpm} \end{matrix} = \dfrac{14,050,000 \text{ gal}}{(74.6 \text{ hrs}) (60 \text{ min/hr})}$

   $= 3139 \text{ gpm}$

   Then determine the filtration rate:

   $\begin{matrix} \text{Filtration} \\ \text{Rate} \end{matrix} = \dfrac{\text{Flow Rate, gpm}}{\text{Area, sq ft}}$

   $= \dfrac{3139 \text{ gpm}}{(35 \text{ ft}) (25 \text{ ft})}$

   $= 3.59 \text{ gpm/sq ft}$

8. $\begin{matrix} \text{Backwash} \\ \text{Rate} \end{matrix} = \dfrac{3120 \text{ gpm}}{375 \text{ sq ft}}$

   $= 8.3 \text{ gpm/sq ft}$

9. $\begin{matrix} \text{Backwash Water} \\ \text{Vol., gal} \end{matrix} = (6,100 \text{ gpm})(6 \text{ min})$

   $= 36,600 \text{ gal}$

10. First calculate the gpm flow rate through the filter, using the $Q = AV$ equation: (14 in. $\div$ 12 in./min = 1.2 ft)

   $Q_{\text{gpm}} = \underset{\text{ft}}{(\text{Length})} \; \underset{\text{ft}}{(\text{Width})} \; \underset{\text{ft/min}}{(\text{Drop Veloc.})} \; \underset{\text{gal/cu ft}}{(7.48)}$

   $= (40 \text{ ft}) (25 \text{ ft}) \left(\dfrac{1.2 \text{ ft}}{5 \text{ min}}\right) \left(7.48 \dfrac{\text{gal}}{\text{cu ft}}\right)$

   $= 1795 \text{ gpm}$

   Then calculate filtration rate:

   $\begin{matrix} \text{Flow Rate,} \\ \text{gpm/sq ft} \end{matrix} = \dfrac{1795 \text{ gpm}}{(40 \text{ hrs}) (25 \text{ ft})}$

   $= 1.8 \text{ gpm/sq ft}$

CHAPTER 11 ACHIEVEMENT TEST—Cont'd

11.   Vol., gal = (0.785) (D ^{2}) (depth) (7.48 gal/cu ft)

$\quad$ 51,600 gal = (0.785) (50 ft) (50 ft) (x ft) (7.48 gal/cu ft)

$$\frac{51,600}{(0.785)\,(50)\,(50)\,(7.48)} = x$$

$$3.5\text{ ft} = x$$

12.   UFRV = (Filtration Rate,) (Filter Run Time,)
$\qquad\qquad\qquad$ gpm/sq ft $\qquad\qquad\qquad$ min

$\qquad$ = (3.1 gpm/sq ft) (3510 min)

$\qquad$ = 10,881 gal/sq ft

13.   First calculate the gpm flow rate, using the Q = AV equation:
(22 in. ÷ 12 in./min = 1.8 ft)

$\qquad$ Q$_{gpm}$ = (Length) (Width) (Drop Veloc.) (7.48)
$\qquad\qquad\qquad$ ft $\qquad$ ft $\qquad$ ft/min $\qquad$ gal/cu ft

$$\qquad = (40\text{ ft}) (20\text{ ft}) \frac{(1.8\text{ ft})}{5\text{ min}} \frac{(7.48\text{ gal})}{\text{cu ft}}$$

$\qquad$ = 2154 gpm

Then calculate the filtration rate:

$$\text{Flow Rate,} \atop \text{gpm/sq ft} \quad = \quad \frac{2154\text{ gpm}}{(40\text{ ft}) (20\text{ ft})}$$

$$= 2.7\text{ gpm/sq ft}$$

14.   Filter Backwash $\quad = \dfrac{3600\text{ gpm}}{(30\text{ ft}) (20\text{ ft})}$
$\quad$ Rate, gpm/sq ft

$$= 6\text{ gpm/sq ft}$$

15.   Backwash Water $\quad$ = (4,700 gpm)(8 min)
$\qquad$ Vol., gal
$\qquad\qquad\qquad\qquad$ = 37,600 gal

16.   Backwash Pumping $\quad$ = (15 gpm/sq ft) (35 ft) (25 ft)
$\qquad$ Rate, gpm

$\qquad\qquad\qquad\qquad$ = 13,125 gpm

17. First calculate the backwash rate as gpm/sq ft:

$$\text{Backwash Rate} \atop \text{gpm/sq ft} \quad = \quad \frac{2900 \text{ gpm}}{(20 \text{ ft})(20 \text{ ft})}$$

$$= 7.3 \text{ gpm/sq ft}$$

Then convert gpm/sq ft to in./min rise rate:

$$\frac{(7.3 \text{ gpm/sq ft}) (12 \text{ in./ft})}{7.48 \text{ gal/cu ft}} = 11.7 \text{ in./min}$$

18. First calculate the gpm flow rate, using the Q = AV equation:
(19 in. ÷ 12 in./min = 1.6 ft)

$$Q_{gpm} = \underset{\text{ft}}{(\text{Length})} \; \underset{\text{ft}}{(\text{Width})} \; \underset{\text{ft/min}}{(\text{Drop Veloc.})} \; \underset{\text{gal/cu ft}}{(7.48)}$$

$$= (35 \text{ ft}) (25 \text{ ft}) \frac{(1.6 \text{ ft})}{7 \text{ min}} (7.48 \frac{\text{gal}}{\text{cu ft}})$$

$$= 1496 \text{ gpm}$$

Then calculate the filtration rate:

$$\text{Flow Rate,} \atop \text{gpm/sq ft} \quad = \quad \frac{1496 \text{ gpm}}{(35 \text{ ft}) (25 \text{ ft})}$$

$$= 1.7 \text{ gpm/sq ft}$$

19. $$\text{Backwash Pumping} \atop \text{Rate, gpm} \quad = (20 \text{ gpm/sq ft}) (40 \text{ ft}) (20 \text{ ft})$$

$$= 16,000 \text{ gpm}$$

20. $$\text{Backwash} \atop \text{Water, \%} \quad = \quad \frac{69,200 \text{ gal}}{17,650,000 \text{ gal}} \times 100$$

$$= 0.4\% \text{ Backwash} \atop \text{Water}$$

21. $$\text{Vol., gal} = (0.785) (D^2) (\text{depth}) (7.48 \text{ gal/cu ft})$$

$$85,200 \text{ gal} = (0.785) (40 \text{ ft}) (40 \text{ ft}) (x \text{ ft}) (7.48 \text{ gal/cu ft})$$

$$\frac{85,200}{(0.785) (40) (40) (7.48)} = 9.1 \text{ ft}$$

CHAPTER 11 ACHIEVEMENT TEST—Cont'd

22. First calculate the volume of mud balls in the sample:

    $$521 \text{ m}L - 500 \text{ m}L \; = \; 21 \text{ m}L$$

    $$\text{\% Mud Ball Volume} = \frac{\text{Mud Ball Vol., m}L}{\text{Total Sample Vol., m}L} \times 100$$

    $$= \frac{21 \text{ m}L}{(5)(695 \text{ m}L)} \times 100$$

    $$= 0.6\%$$

23. $$\text{Backwash Water, \%} = \frac{49{,}100 \text{ gal}}{13{,}700{,}000 \text{ gal}} \times 100$$

    $$= 0.4\%$$

24. First calculate the volume of mud balls in the sample:

    $$568 \text{ m}L - 500 \text{ m}L \; = \; 68 \text{ m}L$$

    $$\text{\% Mud Ball Volume} = \frac{\text{Mud Ball Vol., m}L}{\text{Total Sample Vol., m}L} \times 100$$

    $$= \frac{68 \text{ m}L}{(5)(695 \text{ m}L)} \times 100$$

    $$= 2.0\%$$

Chapter 12

PRACTICE PROBLEMS 12.1

1. $(1.7 \text{ mg}/L) \, (3.2 \text{ MGD}) \, (8.34 \text{ lbs/gal}) \; = \; 45.4 \text{ lbs/day Cl}_2$

2. $(2.4 \text{ mg}/L) \, (1.26 \text{ MGD}) \, (8.34 \text{ lbs/gal}) \; = \; 25.2 \text{ lbs/day}$

3. $\text{Vol., gal} \; = \; (0.785) \, (0.83 \text{ ft}) \, (0.83 \text{ ft}) \, (900 \text{ ft}) \, (7.48 \text{ gal/cu ft})$

    $$= \; 3641 \text{ gal}$$

    Thn calculate the lbs chlorine required using the mg/L to lbs equation:

    $$(\text{mg}/L \text{ Cl}_2) \, (\text{MG Vol.}) \, (8.34 \text{ lbs/gal}) = \text{ lbs Cl}_2$$

    $$(50 \text{ mg}/L) \, (0.003641 \text{ MG}) \, (8.34 \text{ lbs/gal}) = \; 1.5 \text{ lbs Cl}_2$$

4. $(\text{mg/}L\ \text{Cl}_2)\ (\text{MGD flow})\ (8.34\ \text{lbs/gal}) = \text{lbs/day Cl}_2$

$(x\ \text{mg/}L)\ (3.02\ \text{MGD})\ (8.34\ \text{lbs/gal}) = 41\ \text{lbs/day}$

$$x = \frac{41}{(3.02)\ (8.34)}$$

$$x = 1.6\ \text{mg/}L$$

5. First calculate the flow treated:

$$\begin{array}{r} 19{,}414{,}522\ \text{gal} \\ -\ \underline{18{,}762{,}102\ \text{gal}} \\ 652{,}420\ \text{gal/24 hrs} = 0.652\ \text{MGD} \end{array}$$

Now the mg/L chlorine dosage can be calculated:

$(\text{mg/}L\ \text{Cl}_2)\ (\text{MGD flow})\ (8.34\ \text{lbs/gal}) = \text{lbs/day Cl}_2$

$(x\ \text{mg/}L)\ (0.652\ \text{MGD})\ (8.34\ \text{lbs/gal}) = 15\ \text{lbs/day}$

$$x = \frac{15}{(0.652)\ (8.34)}$$

$$x = 2.8\ \text{mg/}L$$

PRACTICE PROBLEMS 12.2

1. Dose = Demand + Residual

$$= 1.4\ \text{mg/}L + 0.5\ \text{mg/}L$$

$$= 1.9\ \text{mg/}L$$

2. Dose = Demand + Residual

$$2.7\ \text{mg/}L = x\ \text{mg/}L + 0.8\ \text{mg/}L$$

$$2.7 - 0.8 = x\ \text{mg/}L$$

$$1.9\ \text{mg/}L = x$$

3. Dose = Demand + Residual

$$= 2.4\ \text{mg/}L + 0.7\ \text{mg/}L$$

$$= 3.1\ \text{mg/}L$$

Then calculate the lbs/day chlorine dose:

$$(3.1\ \text{mg/}L)(3.79\ \text{MGD})(8.34\ \text{lbs/gal}) = 98\ \text{lbs/day}$$

PRACTICE PROBLEMS 12.2—Cont'd

4. First calculate the expected increase in chlorine residual using the mg/L to lbs/day equation:

$$
\underset{\substack{\text{Expected}\\\text{Increase}}}{(\text{mg}/L)\,(\text{MGD flow})\,(8.34\ \text{lbs/gal})} = \underset{\substack{\text{Increase in}\\Cl_2\ \text{Dose}}}{\text{lbs/day}}
$$

$$
(x\ \text{mg}/L)\,(0.96\ \text{MGD})\,(8.34\ \text{lbs/gal}) = 5\ \text{lbs/day}
$$

$$
x = \frac{5}{(0.96)\,(8.34)}
$$

$$
x = 0.6\ \text{mg}/L
$$

The actual increase in residual was:

$$
0.7\ \text{mg}/L - 0.4\ \text{mg}/L = 0.3\ \text{mg}/L
$$

The expected increase in residual was 0.6 mg/L whereas the actual increase in residual was only 0.3 mg/L. From this analysis it appears that the water <u>is not being chlorinated beyond the breakpoint</u>.

5. First calculate the expected increase in chlorine residual:

$$
\underset{\substack{\text{Expected}\\\text{Increase}}}{(\text{mg}/L)\,(\text{MGD flow})\,(8.34\ \text{lbs/gal})} = \underset{\substack{\text{Increase in}\\Cl_2\ \text{Dose}}}{\text{lbs/day}}
$$

$$
(x\ \text{mg}/L)\,(1.98\ \text{MGD})\,(8.34\ \text{lbs/gal}) = 4\ \text{lbs/day}
$$

$$
x = \frac{4}{(1.98)\,(8.34)}
$$

$$
x = 0.2\ \text{mg}/L
$$

The actual increase in chlorine residual was:

$$
0.5\ \text{mg}/L - 0.3\ \text{mg}/L = 0.2\ \text{mg}/L
$$

Since the expected chlorine residual increase (0.2 mg/L) is consistent with the actual increase in chlorine residual (0.2 mg/L), it appears as though the water is being chlorinated beyond the breakpoint.

PRACTICE PROBLEMS 12.3

1.
$$
\underset{\text{lbs/day}}{\text{Hypochlorite}} = \frac{\text{lbs/day } Cl_2}{\dfrac{\%\ \text{Available } Cl_2}{100}}
$$

$$
= \frac{46\ \text{lbs/day}}{0.65}
$$

$$
= 71\ \text{lbs/day Hypochlorite}
$$

2. Hypochlorite
    lbs/day $= \dfrac{\text{lbs/day Cl}_2}{\dfrac{\text{\% Available Cl}_2}{100}}$

    $= \dfrac{38 \text{ lbs/day}}{0.65}$

    $= 58$ lbs/day Hypochlorite

3. First calculate the lbs/day chlorine required:

    $(2.6 \text{ mg/}L)\ (0.917 \text{ MGD})\ (8.34 \text{ lbs/gal}) = 20$ lbs/day Chlorine

    Then calculate the lbs/day hypochlorite required:

    Hypochlorite
    lbs/day $= \dfrac{\text{lbs/day Cl}_2}{\dfrac{\text{\% Available Cl}_2}{100}}$

    $= \dfrac{20 \text{ lbs/day}}{0.65}$

    $= 31$ lbs/day Hypochlorite

4. First calculate the lbs/day chlorine used:

    Hypochlorite
    lbs/day $= \dfrac{\text{lbs/day Cl}_2}{\dfrac{\text{\% Available Cl}_2}{100}}$

    $51 \text{ lbs/day} = \dfrac{x \text{ lbs/day}}{0.65}$

    $(51)\ (0.65) = x$

    33.2 lbs/day $= x$
    chlorine

    Then calculate the mg/L chlorine dosage using the mg/L to lbs/day equation:

    $(x \text{ mg/}L)\ (1.402 \text{ MGD})\ (8.34 \text{ lbs/gal}) = 33.2$ lbs/day

    $$x = \dfrac{33.2}{(1.402)\ (8.34)}$$

    $$x = 2.8 \text{ mg/}L$$

PRACTICE PROBLEMS 12.3—Cont'd

5. To determine the mg/*L* chlorine dosage, you must first determine the lbs/day chlorine dosage:

$$\frac{\text{Hypochlorite}}{\text{lbs/day}} = \frac{\text{lbs/day Cl}_2}{\dfrac{\% \text{ Available Cl}_2}{100}}$$

$$47 \text{ lbs/day} = \frac{x \text{ lbs/day}}{0.70}$$

$$(47)(0.70) = x$$

$$\underset{\text{chlorine}}{32.9 \text{ lbs/day}} = x$$

Then calculate the mg/*L* chlorine dosage:

$$(x \text{ mg/}L)(2.944 \text{ MGD})(8.34 \text{ lbs/gal}) = 32.9 \text{ lbs/day}$$

$$x = \frac{32.9}{(2.944)(8.34)}$$

$$x = 1.3 \text{ mg/}L \text{ Chlorine}$$

PRACTICE PROBLEMS 12.4

1. $\dfrac{35 \text{ lbs/day}}{8.34 \text{ lbs/gal}} = 4.2 \text{ gpd Hypochlorite}$

2. $\underset{\text{Dose}}{(\text{mg/}L)} \underset{\text{Treated}}{(\text{MGD Flow})} \underset{\text{lbs/gal}}{(8.34)} = \underset{\text{Sol'n}}{(\text{mg/}L)} \underset{\text{Sol'n}}{(\text{MGD})} \underset{\text{lbs/gal}}{(8.34)}$

$$\underset{\text{lbs/gal}}{(2.9 \text{ mg/}L)(0.795 \text{ MGD})(8.34)} = \underset{\text{lbs/gal}}{(130,000 \text{ mg/}L)(x \text{ MGD})(8.34)}$$

$$\frac{(2.9)(0.795)(8.34)}{(130,000)(8.34)} = x$$

$$0.0000177 \text{ MGD} = x$$

$$17.7 \text{ gpd} = x$$

3. $\underset{\text{Dose}}{(\text{mg/}L)} \underset{\text{Treated}}{(\text{MGD Flow})} \underset{\text{lbs/gal}}{(8.34)} = \underset{\text{Sol'n}}{(\text{mg/}L)} \underset{\text{Sol'n}}{(\text{MGD})} \underset{\text{lbs/gal}}{(8.34)}$

$$\underset{\text{lbs/gal}}{(2.1 \text{ mg/}L)(0.22 \text{ MGD})(8.34)} = \underset{\text{lbs/gal}}{(20,000 \text{ mg/}L)(x \text{ MGD})(8.34)}$$

$$\frac{(2.1)(0.22)(8.34)}{(20,000)(8.34)} = x$$

$$0.0000231 \text{ MGD} = x$$

$$23.1 \text{ gpd} = x$$

4. The water flow rate and solution flow rate must be expressed as gpd (then MGD) for use in the equation.

   <u>Well Water Flow Rate:</u>

   $$\frac{2{,}110{,}000 \text{ gal}}{7 \text{ days}} = 301{,}429 \text{ gpd}$$

   <u>Solution Feeder Flow Rate:</u>

   $$\frac{(0.785)(3 \text{ ft})(3 \text{ ft})(2.75 \text{ ft})(7.48 \text{ gal/cu ft})}{7 \text{ days}} = 21 \text{ gpd}$$

   Now the mg/L chlorine dose can be calculated:

   (mg/L) (MGD Flow) (8.34) = (mg/L) (MGD) (8.34)
     Dose   Treated lbs/gal Sol'n Sol'n lbs/gal

   (x mg/L)(0.301)(8.34) = (20,000 mg/L)(0.000021)(8.34)
           MGD lbs/gal   MGD lbs/gal

   $$x = \frac{(20{,}000)(0.000021)(\cancel{8.34})}{(0.301)(\cancel{8.34})}$$

   $$x = 1.4 \text{ mg/}L$$

5. First express the flow rate to be treated as MGD:

   $$(300 \text{ gpm})(1440 \text{ min/day}) = 432{,}000 \text{ gpd}$$

   $$\text{or} = 0.432 \text{ MGD}$$

   Then calculate the solution feed rate:

   (mg/L) (MGD Flow) (8.34) = (mg/L) (MGD) (8.34)
     Dose   Treated lbs/gal Sol'n Sol'n lbs/gal

   (1.6 mg/L)(0.432 MGD)(8.34) = (30,000 mg/L)(x MGD)(8.34)
               lbs/gal   lbs/gal

   $$\frac{(1.6)(0.432)(\cancel{8.34})}{(30{,}000)(\cancel{8.34})} = x$$

   $$0.000023 \text{ MGD} = x$$

   $$23 \text{ gpd} = x$$

PRACTICE PROBLEMS 12.4—Cont'd

6. $\underset{\text{Dose}}{(mg/L)} \underset{\text{Treated}}{(MGD \ Flow)} \underset{\text{lbs/gal}}{(8.34)} = \underset{\text{Sol'n}}{(mg/L)} \underset{\text{Sol'n}}{(MGD)} \underset{\text{lbs/gal}}{(8.34)}$

$\underset{\text{lbs/gal}}{(2.7 \ mg/L)(0.94 \ MGD)(8.34)} = \underset{\text{lbs/gal}}{(20,000 \ mg/L)(x \ MGD)(8.34)}$

$$\frac{(2.7)(0.94)\cancel{(8.34)}}{(20,000)\cancel{(8.34)}} = x$$

$$0.0001269 \ MGD = x$$

$$126.9 \ gpd = x$$

PRACTICE PROBLEMS 12.5

1.
$$\underset{\text{Strength}}{\% \ Cl_2} = \frac{\dfrac{(\text{Hypochl., lbs})(\% \ \text{Avail. } Cl_2)}{100}}{\text{Water, lbs} + \dfrac{(\text{Hypochl., lbs})(\% \ \text{Avail. } Cl_2)}{100}} \times 100$$

$$= \frac{(20 \ \text{lbs}) \ (0.65)}{(55 \ \text{gal}) \ (8.34 \ \text{lbs/gal}) + \ (20 \ \text{lbs}) \ (0.65)} \times 100$$

$$= \frac{13 \ \text{lbs}}{458.7 \ \text{lbs} + 13 \ \text{lbs}} \times 100$$

$$= \frac{13 \ \text{lbs}}{471.7 \ \text{lbs}}$$

$$= 2.8\% \ \text{chlorine}$$

2. First, convert grams to lbs:

$$(300 \ \text{grams}) \ (0.0022 \ \text{lbs/gram}) = 0.66 \ \text{lbs}$$
$$\text{hypochlorite}$$

Then calculate percent strength:

$$\underset{\text{Strength}}{\% \ Cl_2} = \frac{\dfrac{(\text{Hypochl., lbs})(\% \ \text{Avail. } Cl_2)}{100}}{\text{Water, lbs} + \dfrac{(\text{Hypochl., lbs})(\% \ \text{Avail. } Cl_2)}{100}} \times 100$$

$$= \frac{(0.66 \ \text{lbs}) \ (0.65)}{(5 \ \text{gal}) \ (8.34 \ \text{lbs/gal}) + \ (0.66 \ \text{lbs}) \ (0.65)} \times 100$$

$$= \frac{0.4 \ \text{lbs}}{41.7 \ \text{lbs} + 0.4 \ \text{lbs}} \times 100$$

$$= \frac{0.4 \ \text{lbs}}{42.1 \ \text{lbs}}$$

$$= 1.0\% \ \text{chlorine}$$

3. $\underset{\text{Strength}}{\% \text{ Cl}_2} = \cfrac{\cfrac{(\text{Hypochl., lbs})(\% \text{ Avail. Cl}_2)}{100}}{\text{Water, lbs} + \cfrac{(\text{Hypochl., lbs})(\% \text{ Avail. Cl}_2)}{100}} \times 100$

$2 = \cfrac{(x \text{ lbs}) (0.70)}{(55 \text{ gal}) (8.34 \text{ lbs/gal}) + (x \text{ lbs}) (0.70)} \times 100$

$2 = \cfrac{(x) (0.70) (100)}{458.7 + 0.70 \, x}$

$2 = \cfrac{70 \, x}{458.7 + 0.70 \, x}$

$458.7 + 0.70 \, x = 35 \, x$

$458.7 = 34.3 \, x$

$\cfrac{458.7}{34.3} = x$

$13.4 \text{ lbs} = x$

4. $\underset{\text{gal} \quad \text{lbs/gal}}{(\text{Liq. Hypo.}) (8.34)} \ \underset{100}{\dfrac{(\% \text{ Avail.})}{\text{Chlorine}}} = \underset{\text{gal} \quad \text{lbs/gal}}{(\text{Hypo. Sol'n}) (8.34)} \ \underset{100}{\dfrac{(\% \text{ Avail.})}{\text{Chlorine}}}$

$(x \text{ gal}) (8.34) \dfrac{(12)}{100} = (30 \text{ gal}) (8.34) \dfrac{(2)}{100}$

$x = \dfrac{(30) (8.34) (0.02)}{(8.34) (0.12)}$

$x = 5 \text{ gal}$

5. $\underset{\text{gal} \quad \text{lbs/gal}}{(\text{Liq. Hypo.}) (8.34)} \ \underset{100}{\dfrac{(\% \text{ Avail.})}{\text{Chlorine}}} = \underset{\text{gal} \quad \text{lbs/gal}}{(\text{Hypo. Sol'n}) (8.34)} \ \underset{100}{\dfrac{(\% \text{ Avail.})}{\text{Chlorine}}}$

$(x \text{ gal}) (8.34) \dfrac{(14)}{100} = (100 \text{ gal}) (8.34) \dfrac{(1.3)}{100}$

$x = \dfrac{(100) (8.34) (0.013)}{(8.34) (0.14)}$

$x = 9.3 \text{ gal}$

PRACTICE PROBLEMS 12.5—Cont'd

6. $\underset{\text{gal}}{\text{(Liq. Hypo.)}} \underset{\text{lbs/gal}}{\text{(8.34)}} \underset{\text{100}}{\text{(% Avail.)}\atop\underline{\text{Chlorine}}} = \underset{\text{gal}}{\text{(Hypo. Sol'n)}} \underset{\text{lbs/gal}}{\text{(8.34)}} \underset{\text{100}}{\text{(% Avail.)}\atop\underline{\text{Chlorine}}}$

$$(5 \text{ gal}) (8.34) \left(\frac{13}{100}\right) = (x \text{ gal}) (8.34) \left(\frac{3}{100}\right)$$

$$\frac{(5)\,(8.34)\,(0.13)}{(8.34)\,(0.03)} = x$$

$$\underset{\text{Solution}}{21.7 \text{ gal}} = x$$

The solution is comprised of a total of 21.7 gallons solution. Since 5 gallons are liquid hypochlorite, a total of 21.7 gal − 5 gal = 16.7 gal water must be added to the drum.

PRACTICE PROBLEMS 12.6

1. $\underset{\text{of Mixture}}{\% \text{ Cl}_2 \atop \text{Strength}} = \dfrac{\dfrac{(\text{Sol'n 1})}{\text{lbs}} \dfrac{(\% \text{ Avail. Cl}_2)}{\text{of Sol'n 1}}}{100} + \dfrac{(\text{Sol'n 2})}{\text{lbs}} \dfrac{(\% \text{ Avail. Cl}_2)}{\text{of Sol'n 2}}}{\text{lbs Solution 1 + lbs Solution 2}} \times 100$

$$= \frac{(40 \text{ lbs})\,(0.12) + (250 \text{ lbs})\,(0.01)}{40 \text{ lbs} + 250 \text{ lbs}} \times 100$$

$$= \frac{4.8 \text{ lbs} + 2.5 \text{ lbs}}{290 \text{ lbs}} \times 100$$

$$= \frac{7.3 \text{ lbs}}{290 \text{ lbs}} \times 100$$

$$= 2.5\%$$

2. $\underset{\text{of Mixture}}{\% \text{ Cl}_2 \atop \text{Strength}} = \dfrac{\dfrac{(\text{Sol'n 1})}{\text{gal}}\dfrac{(8.34)}{\text{lbs/gal}}\dfrac{(\% \text{ Avail. Cl}_2)}{\text{of Sol'n 1}}}{100} + \dfrac{(\text{Sol'n 2})}{\text{gal}}\dfrac{(8.34)}{\text{lbs/gal}}\dfrac{(\% \text{ Avail. Cl}_2)}{\text{of Sol'n 2}}}{100}}{\text{Solution 1, gal + Solution 2, gal}} \times 100$

$$= \frac{(8 \text{ gal})\,(8.34 \text{ lbs/gal})\left(\dfrac{10}{100}\right) + (50 \text{ gal})\,(8.34 \text{ lbs/gal})\left(\dfrac{1.5}{100}\right)}{(8 \text{ gal})\,(8.34 \text{ lbs/gal}) + (50 \text{ gal})\,(8.34 \text{ lbs/gal})} \times 100$$

$$= \frac{6.7 \text{ lbs} + 6.3 \text{ lbs}}{66.7 \text{ lbs} + 417 \text{ lbs}} \times 100$$

$$= \frac{13 \text{ lbs}}{483.7 \text{ lbs}} \times 100$$

$$= 2.7\%$$

3. $\%\ Cl_2$ Strength of Mixture $=$

$$\dfrac{\dfrac{(\text{Sol'n 1})(8.34)}{\text{gal}\quad\text{lbs/gal}}\dfrac{(\%\ \text{Avail. }Cl_2)}{\text{of Sol'n 1}}}{100} + \dfrac{(\text{Sol'n 2})(8.34)}{\text{gal}\quad\text{lbs/gal}}\dfrac{(\%\ \text{Avail. }Cl_2)}{\text{of Sol'n 2}}}{100}}{\text{Solution 1, gal} + \text{Solution 2, gal}} \times 100$$

$$= \dfrac{(15\text{ gal})\,(8.34\text{ lbs/gal})\dfrac{(11)}{100} + (65\text{ gal})\,(8.34\text{ lbs/gal})\dfrac{(1)}{100}}{(15\text{ gal})\,(8.34\text{ lbs/gal}) + (65\text{ gal})\,(8.34\text{ lbs/gal})} \times 100$$

$$= \dfrac{13.8\text{ lbs} + 5.4\text{ lbs}}{125.1\text{ lbs} + 542.1\text{ lbs}} \times 100$$

$$= \dfrac{19.2\text{ lbs}}{667.2\text{ lbs}} \times 100$$

$$= 2.9\%$$

4. Use the Dilution Rectangle to solve this problem:

13 ⟍ ⟶ 1.5
    3
1.5 ⟋ ⟶ 10

11.5 parts

Amt. of 13% Soln: $\dfrac{(1.5)}{11.5}\,(800\text{ lbs}) = 104.3\text{ lbs}$

Amt. of 1.5% Soln: $\dfrac{(10)}{11.5}\,(800\text{ lbs}) = 695.7\text{ lbs}$

5.
14 ⟍ ⟶ 1
    2
1 ⟋ ⟶ 12

13 parts

Amt. of 14% Soln: $\dfrac{(1)}{13}\,(500\text{ lbs}) = 38.5\text{ lbs}$

Amt. of 1% Soln: $\dfrac{(12)}{13}\,(500\text{ lbs}) = 461.5\text{ lbs}$

6. Since the densities of the solutions are assumed to be the same, the Dilution Rectangle method can be used to determine **gallons** of each solution required: (Remember, water is considered a 0% strength solution.)

11 ⟍ ⟶ 2
    2
0 ⟋ ⟶ 9

11 parts

Amt. of 11% Soln: $\dfrac{(2)}{11}\,(400\text{ gal}) = 72.7\text{ gal}$

Amt. of Water: $\dfrac{(9)}{11}\,(400\text{ gal}) = 327.3\text{ gal}$

PRACTICE PROBLEMS 12.7

1. $\begin{array}{l} \text{Days' Supply} \\ \text{in Inventory} \end{array} = \dfrac{\text{Total Chem. in Inventory, lbs}}{\text{Average Use, lbs/day}}$

    $= \dfrac{900\ \text{lbs}}{41\ \text{lbs/day}}$

    $= 22\ \text{days}$

2. (7 in. ÷ 12 in./ft = 0.58 ft)

    $\begin{array}{l} \text{Days' Supply} \\ \text{in Tank} \end{array} = \dfrac{\text{Total Solution in Tank, gal}}{\text{Average Use, gpd}}$

    $x\ \text{days} = \dfrac{(0.785)\ (3\ \text{ft})\ (3\ \text{ft})\ (3.58\ \text{ft})\ (7.48\ \text{gal/cu ft})}{70\ \text{gpd}}$

    $= 2.7\ \text{days' Supply}$

3. $\begin{array}{l} \text{Total Chem.} \\ \text{Used, lbs} \end{array} = \dfrac{(\text{Chem. Use,})\ (\text{Hrs Used})}{\text{lbs/hr}}$

    $= \dfrac{(26\ \text{lbs/day})\ (140\ \text{hrs operation})}{24\ \text{hrs/day}}$

    $= 152\ \text{lbs chlorine used}$

4. First calculate the lbs chlorine used:

    $\begin{array}{l} \text{Total Chem.} \\ \text{Used, lbs} \end{array} = \dfrac{(\text{Chem. Use,})\ (\text{Hrs Used})}{\text{lbs/hr}}$

    $= \dfrac{(14\ \text{lbs/day})\ (105\ \text{hrs operation})}{24\ \text{hrs/day}}$

    $= 61.3\ \text{lbs}$

    Then calculate the lbs chlorine remaining at the end of the week:

    89 lbs – 61.3 lbs = 27.7 lbs remaining

5. The chlorine used during the month is:

    $(52\ \dfrac{\text{lbs}}{\text{day}})\ (30\ \text{days}) = 1560\ \text{lbs chlorine/month}$

    The number of 150-lb containers required for the month is:

    $\dfrac{1560\ \text{lbs Cl}_2}{150\ \text{lbs/cylinder}} = 10.4\ \text{cylinders}$
    (11 cylinders would be required)

6. Days' Supply $= \dfrac{\text{Total Solution in Tank, gal}}{\text{Average Use, gpd}}$

$2 \text{ days} = \dfrac{(0.785)\ (3\ \text{ft})\ (3\ \text{ft})\ (x\ \text{ft})\ (7.48\ \text{gal/cu ft})}{60\ \text{gpd}}$

$\dfrac{(2)\ (60)}{(0.785)(3)(3)(7.48)} = x$

$2.3 \text{ ft} = x$

CHAPTER 12—ACHIEVEMENT TEST

1. Dose = Demand + Residual

   $= 1.6 \text{ mg/}L + 0.8 \text{ mg/}L$

   $= 2.4 \text{ mg/}L$

2. $(2.2 \text{ mg/}L)(0.91 \text{ MGD})(8.34 \text{ lbs/gal}) = 17 \text{ lbs/day}$

3. $\text{Hypochlorite, lbs/day} = \dfrac{\text{lbs/day Cl}_2}{\dfrac{\%\ \text{Available Cl}_2}{100}}$

   $= \dfrac{50 \text{ lbs/day}}{0.65}$

   $= 76.9 \text{ lbs/day Hypochlorite}$

4. $\dfrac{48 \text{ lbs/day}}{8.34 \text{ lbs/gal}} = 5.8 \text{ gpd}$

5. Dose = Demand + Residual

   $2.9 \text{ mg/}L = x \text{ mg/}L + 0.6 \text{ mg/}L$

   $2.9 - 0.6 = x$

   $2.3 \text{ mg/}L = x$

CHAPTER 12—ACHIEVEMENT TEST—Cont'd

6. $\dfrac{\% \, Cl_2}{\text{Strength}} = \dfrac{\text{lbs Chlorine}}{\text{lbs Solution}} \times 100$

$= \dfrac{(25 \text{ lbs}) \, (0.65)}{(62 \text{ gal}) \, (8.34 \text{ lbs/gal}) + (25 \text{ lbs}) \, (0.65)} \times 100$

$= \dfrac{16.3 \text{ lbs}}{517 \text{ lbs} + 16.3 \text{ lbs}} \times 100$

$= 3\%$

7. $(1580 \text{ gpm}) \, (1440 \text{ min/day}) = 2{,}275{,}200 \text{ gpd}$

$(2.6 \text{ mg/L}) \, (2.275 \text{ MGD}) \, (8.34 \text{ lbs/gal}) = 49.3 \text{ lbs/day}$

8. Actual Dose. lbs/day = Solution Feeder Dose, lbs/day

$\underset{\text{lbs/gal}}{(2.6 \text{ mg}/L)(1.17 \text{ MGD})(8.34)} = \underset{\text{lbs/gal}}{(12{,}000 \text{ mg}/L)(x \text{ MGD})(8.34)}$

$\dfrac{(2.6)(1.17)\cancel{(8.34)}}{(12{,}000)\cancel{(8.34)}} = x$

$0.0002535 \text{ MGD} = x$

$253.5 \text{ gpd} = x$

9. First calculate the volume of water to be treated:

Vol. gal $= (0.785) \, (0.5 \text{ ft}) \, (0.5 \text{ ft}) \, (1500 \text{ ft}) \, (7.48 \text{ gal/cu ft})$

$= 2202 \text{ gal}$

Then calculate the lbs Cl_2:

$(\text{mg}/L \, Cl_2) \, (\text{MG Vol.}) \, (8.34 \text{ lbs/gal}) = \text{lbs } Cl_2$

$(50 \text{ mg}/L) \, (0.002202 \text{ MG}) \, (8.34 \text{ lbs/gal}) = 0.9 \text{ lbs}$

10. First, calculate the expected increase in chlorine residual:

$$\begin{array}{l}\text{(mg/}L\text{) (MGD flow) (8.34 lbs/gal)} \\ \text{Expected} \\ \text{Increase}\end{array} = \begin{array}{l}\text{lbs/day} \\ \text{Increase in} \\ \text{Cl}_2 \text{ Dose}\end{array}$$

$$(x \text{ mg/}L) \ (2.07 \text{ MGD}) \ (8.34 \text{ lbs/gal}) = 4 \text{ lbs/day}$$

$$x = \frac{4}{(2.07) \ (8.34)}$$

$$x = 0.23 \text{ mg/}L$$

The actual increase in residual was:

$$0.7 \text{ mg/}L - 0.5 \text{ mg/}L = 0.2 \text{ mg/}L$$

The expected and actual residual increases are approximately the same. Therefore chlorination is assumed to be beyond the breakpoint.

11.
$$\begin{array}{l}\% \ \text{Cl}_2 \\ \text{Strength}\end{array} = \frac{\begin{array}{l}\text{lbs Cl}_2 \text{ in} \\ \text{Sol'n 1}\end{array} + \begin{array}{l}\text{lbs Cl}_2 \text{ in} \\ \text{Sol'n 2}\end{array}}{\text{lbs Sol'n 1} + \text{lbs Sol'n 2}} \times 100$$

$$= \frac{(60 \text{ gal}) \ (8.34 \text{ lbs/gal}) \ \dfrac{(12)}{100} + (240 \text{ gal}) \ (8.34 \text{ lbs/gal}) \ \dfrac{(2)}{100}}{(60 \text{ gal}) \ (8.34 \text{ lbs/gal}) + (240 \text{ gal}) \ (8.34 \text{ lbs/gal})} \times 100$$

$$= \frac{60 \text{ lbs} + 40 \text{ lbs}}{500.4 \text{ lbs} + 2001.6 \text{ lbs}} \times 100$$

$$= \frac{100 \text{ lbs}}{2502 \text{ lbs}} \times 100$$

$$= 4\%$$

12.
$$\text{Days' Supply} = \frac{\text{Total Chem. in Inventory, lbs}}{\text{Average Use, lbs/day}}$$

$$= \frac{280 \text{ lbs}}{36 \text{ lbs/day}}$$

$$= 7.8 \text{ days}$$

CHAPTER 12—ACHIEVEMENT TEST—Cont'd

13. First determine the flow rate treated:

$$
\begin{array}{r}
43,981,669 \text{ gal} \\
- \underline{42,197,660 \text{ gal}} \\
1,784,009 \text{ gal}
\end{array}
$$

Then calculate the mg/L chlorine dosage:

$$(x \text{ mg}/L)\ (1.784 \text{ MGD})\ (8.34 \text{ lbs/gal}) = 19 \text{ lbs/day}$$

$$x = \frac{19}{(1.784)\ (8.34)}$$

$$x = 1.3 \text{ mg}/L$$

14. $\text{Hypochlorite, lbs/day} = \dfrac{\text{lbs/day Cl}_2}{\dfrac{\%\ \text{Available Cl}_2}{100}}$

$$= \frac{35 \text{ lbs/day}}{0.70}$$

$$= 50 \text{ lbs/day Hypochlorite}$$

15. First calculate the flow rate treated and the solution feeder flow rate:

Flow Rate Treated, gpd

$$\frac{2,417,000 \text{ gal}}{7 \text{ days}} = 345,286 \text{ gpd}$$

Solution Feeder Flow Rate, gpd (2 in. ÷ 12 in./ft = 0.17 ft)

$$\frac{(0.785)(3 \text{ ft})(3 \text{ ft})(3.17 \text{ ft})(7.48 \text{ gal/cu ft})}{7 \text{ days}} = 24 \text{ gpd}$$

Now the mg/L chlorine dose can be calculated:

Actual Dose, lbs/day = Solution Feeder Dose, lbs/day

$$\underset{\text{MGD} \quad \text{lbs/gal}}{(x \text{ mg}/L)\ (0.345)\ (8.34)} = \underset{\text{MGD} \quad \text{lbs/gal}}{(20,000 \text{ mg}/L)\ (0.000024)\ (8.34)}$$

$$x = \frac{(20,000)\ (0.000024)\ \cancel{(8.34)}}{(0.345)\ \cancel{(8.34)}}$$

$$x = 1.4 \text{ mg}/L$$

16. Chlorine dose = 2.8 mg/L

    The chlorinator setting should therefore be:

    $$(2.8 \text{ mg}/L) \ (3.15 \text{ MGD}) \ (8.34 \text{ lbs/gal}) \ = \ 73.6 \text{ lbs/day}$$

17. Assuming the densities of the solutions are 8.34 lbs/gal, the equation using gallons rather than pounds may be used:

    $$\frac{\% \ Cl_2}{\text{Strength}} = \frac{\text{lbs } Cl_2 \text{ in Sol'n 1} + \text{lbs } Cl_2 \text{ in Sol'n 2}}{\text{lbs Sol'n 1} + \text{lbs Sol'n 2}} \times 100$$

    $$= \frac{(10 \text{ gal}) \dfrac{(14)}{100} + (40 \text{ gal}) \dfrac{(1)}{100}}{10 \text{ gal} + 40 \text{ gal}} \times 100$$

    $$= \frac{1.4 \text{ gal} + 0.4 \text{ gal}}{50 \text{ gal}} \times 100$$

    $$= \frac{1.8 \text{ gal}}{50 \text{ gal}} \times 100$$

    $$= \ 3.6\%$$

18. First, calculate the lbs/day chlorine used:

    $$\frac{\text{Hypochlorite,}}{\text{lbs/day}} = \frac{\text{lbs/day } Cl_2}{\dfrac{\% \text{ Available } Cl_2}{100}}$$

    $$65 \text{ lbs/day} = \frac{x}{0.65}$$

    $$(65) \ (0.65) = x$$

    $$42.3 \text{ lbs/day} = x$$

    Then calculate the mg/L chlorine dosage:

    $$(x \text{ mg}/L) \ (1.75 \text{ MGD}) \ (8.34 \text{ lbs/gal}) = 42.3 \text{ lbs/day}$$

    $$x = \frac{42.3}{(1.75) \ (8.34)}$$

    $$x = \ 2.9 \text{ mg}/L$$

CHAPTER 12—ACHIEVEMENT TEST—Cont'd

19. First, convert the flow treated to MGD:

$$(350 \text{ gpm})(1440 \text{ min/day}) = 504,000 \text{ gpd}$$
$$\text{or} = 0.504 \text{ MGD}$$

Then calculate the hypochlorite solution flow required:

Actual Dose, lbs/day = Solution Feeder Dose, lbs/day

$$(2.4 \text{ mg}/L)(0.504 \text{ MGD})(8.34 \text{ lbs/gal}) = (30,000 \text{ mg}/L)(x \text{ MGD})(8.34 \text{ lbs/gal})$$

$$\frac{(2.4)(0.504)\cancel{(8.34)}}{(30,000)\cancel{(8.34)}} = x$$

$$0.0000403 \text{ MGD} = x$$

$$40.3 \text{ gpd} = x$$

20. Days' Supply $= \dfrac{\text{Total Chem. in Inventory, lbs}}{\text{Average Use, lbs/day}}$

$$= \frac{(0.785)(3 \text{ ft})(3 \text{ ft})(3.92 \text{ ft})(7.48 \text{ gal/cu ft})}{85 \text{ gpd}}$$

$$= 2.4 \text{ days}$$

21. First, calculate the lbs of chlorine required:

$$\begin{array}{l} \% \text{ Cl}_2 \\ \text{Strength} \end{array} = \frac{\text{Chlorine, lbs}}{\text{Water, lbs} + \text{Chlorine, lbs}} \times 100$$

$$2 = \frac{x \text{ lbs}}{(75 \text{ gal}) (8.34 \text{ lbs/gal}) + x \text{ lbs}} \times 100$$

$$2 = \frac{100\,x}{625.5 + x}$$

$$625.5 + x = \frac{100\,x}{2}$$

$$625.5 + x = 50\,x$$

$$625.5 = 49\,x$$

$$\frac{625.5}{49} = x$$

$$12.8 \text{ lbs} = x$$

Then calculate the lbs hypochlorite required:

$$\text{Hypochlorite, lbs} = \frac{\text{Chlorine, lbs}}{\dfrac{\%\ \text{Available Cl}_2}{100}}$$

$$= \frac{12.8\ \text{lbs}}{0.65}$$

$$= 19.7\ \text{lbs Hypochlorite}$$

22. $\dfrac{28\ \text{lbs/day}}{24\ \text{hrs/day}} = 1.17\ \text{lbs/hr chlorine used}$

For 135 hrs of operation, then, the chlorine use would be:

$$(1.17\ \text{lbs/hr})(135\ \text{hrs}) = 158\ \text{lbs}$$

23. The total chlorine used for 30 days would be:

$$(45\ \text{lbs/day})(30\ \text{days}) = 1350\ \text{lbs}$$

The number of 150-lb cylinders required would be:

$$\frac{1350\ \text{lbs}}{150\ \text{lbs/cylinder}} = 9\ \text{cylinders}$$

24.

12 → 1.5   Amt. of 12% Sol'n: $= \dfrac{(1.5)}{12}$ (200 gal) $= 25$ gal

   1.5

0 → 10.5   Amt. of Water: $= \dfrac{(10.5)}{12}$ (200 gal) $= 175$ gal

   12 parts

Chapter 13

PRACTICE PROBLEMS 13.1

1. $2.4\% = 24{,}000$ mg/*L*

2. 6700 mg/*L* $= 0.67\%$

3. $22\% = 220{,}000$ mg/*L*

4. $\dfrac{14\ \text{lbs}}{(1\ \text{MG})(8.34\ \text{lbs/gal})} = \dfrac{14\ \text{lbs}}{8.34\ \text{mil lbs}} = \dfrac{1.68\ \text{lbs}}{1\ \text{mil lbs}} = 1.68$ mg/*L*

PRACTICE PROBLEMS—Cont'd

5. $1.8 \text{ mg/}L = \dfrac{1.8 \text{ lbs}}{1 \text{ mil lbs}} = \dfrac{1.8 \text{ lbs}}{\dfrac{1 \text{ mil lbs}}{8.34 \text{ lbs/gal}}} = \dfrac{1.8 \text{ lbs}}{0.12 \text{ MG}} = \dfrac{15 \text{ lbs}}{1 \text{ MG}}$

6. $\dfrac{23 \text{ lbs}}{(1 \text{ MG})(8.34 \text{ lbs/gal})} = \dfrac{23 \text{ lbs}}{8.34 \text{ mil lbs}} = \dfrac{2.76 \text{ lbs}}{1 \text{ mil lbs}} = 2.76 \text{ mg/}L$

PRACTICE PROBLEMS 13.2

1. First calculate the molecular weight of each element in the compound:

Element	No. of Atoms		Atomic Wt.		Molec. Wt.
H:	2	x	1.008	=	2.016
Si:	1	x	28.06	=	28.06
F:	6	x	19.00	=	114.00
					144.076

   The percent fluoride in the compound can now be calculated:

   $$\% \text{ F in } H_2SiF_6 = \frac{\text{Molec. Wt. of F}}{\text{Molec. Wt. of } H_2SiF_6} \times 100$$

   $$= \frac{114.00}{144.076} \times 100$$

   $$= 79.1\%$$

2. First calculate the molecular weight of each element in the compound:

Element	No. of Atoms	Atomic Wt.	Molec. Wt.
Na:	1	22.997	22.997
F:	1	19.00	19.00
		Molec. Wt. of NaF:	41.997

   Then calculate the percent fluoride in NaF:

   $$\% \text{ F in NaF} = \frac{\text{Molec. Wt. of F}}{\text{Molec. Wt. of NaF}} \times 100$$

   $$= \frac{19.00}{41.997} \times 100$$

   $$= 45.2\%$$

3. Calculate the molecular weight of each element in the compound:

Element	No. of Atoms		Atomic Wt.		Molec. Wt.
Na:	2	x	22.997	=	45.994
Si:	1	x	28.06	=	28.06
F:	6	x	19.00	=	114.00
					188.054

Then calculate the percent fluoride in NaF:

$$\text{\% F in } Na_2SiF_6 = \frac{\text{Molec. Wt. of F}}{\text{Molec. Wt. of } Na_2SiF_6} \times 100$$

$$= \frac{114.00}{188.054} \times 100$$

$$= 60.6\%$$

4. First calculate the molecular weight of each element in the compound:

Element	No. of Atoms		Atomic Wt.		Molec. Wt.
Mg:	1	x	24.32	=	24.32
Si:	1	x	28.06	=	28.06
F:	6	x	19.00	=	114.00
					166.38

The percent fluoride ion can now be calculated:

$$\text{\% F in } MgSiF_6 = \frac{\text{Molec. Wt. of F}}{\text{Molec. Wt. of } MgSiF_6} \times 100$$

$$= \frac{114.00}{166.38} \times 100$$

$$= 68.5\%$$

PRACTICE PROBLEMS 13.3

1. $\dfrac{(1.5 \text{ mg}/L)\,(0.98 \text{ MG})\,(8.34 \text{ lbs/gal})}{\dfrac{(98)}{100}\,\dfrac{(60.6)}{100}} = \begin{array}{l} 20.6 \text{ lbs/day} \\ Na_2SiF_6 \end{array}$

2. $\dfrac{(1.2 \text{ mg}/L)\,(1.68 \text{ MG})\,(8.34 \text{ lbs/gal})}{\dfrac{(98)}{100}\,\dfrac{(60.6)}{100}} = \begin{array}{l} 28.3 \text{ lbs/day} \\ Na_2SiF_6 \end{array}$

PRACTICE PROBLEMS—Cont'd

3. $\dfrac{(1.6 \text{ mg}/L)\ (2.796 \text{ MGD})\ (8.34 \text{ lbs/gal})}{0.6} = \begin{array}{l} 62.2 \text{ lbs/day} \\ Na_2SiF_6 \end{array}$

4. $\dfrac{(1.0 \text{ mg}/L)\ (3.06 \text{ MGD})\ (8.34 \text{ lbs/gal})}{0.44} = \begin{array}{l} 58 \text{ lbs/day} \\ NaF \end{array}$

5. The fluoride ion to be added to the water is 1.3 mg/L – 0.09 mg/L = 1.21 mg/L

$$\dfrac{(1.21 \text{ mg}/L)\ (0.775 \text{ MGD})\ (8.34 \text{ lbs/gal})}{0.45} = \begin{array}{l} 17.4 \text{ lbs/day} \\ NaF \end{array}$$

PRACTICE PROBLEMS 13.4

1. % Strength $= \dfrac{(\text{lbs Chem.})\ \dfrac{(\%\ \text{Comm. Purity})}{100}}{\text{lbs Water} + (\text{lbs Chem.})\ \dfrac{(\%\ \text{Comm. Pur.})}{100}} \times 100$

$= \dfrac{(8 \text{ lbs})\ \dfrac{(98)}{100}}{(50 \text{ gal})\ (8.34 \text{ lbs/gal}) + (8 \text{ lbs})\ \dfrac{(98)}{100}} \times 100$

$= \dfrac{7.84}{417 + 7.84} \times 100$

$= 1.8\%$ Strength NaF

2. % Strength $= \dfrac{\text{lbs Chemical}}{\text{lbs Water} + \text{lbs Chemical}} \times 100$

$= \dfrac{15 \text{ lbs}}{(75 \text{ gal})\ (8.34 \text{ lbs/gal}) + 15 \text{ lbs}} \times 100$

$= \dfrac{15 \text{ lbs}}{625.5 \text{ lbs} + 15 \text{ lbs}} \times 100$

$= \dfrac{15}{640.5} \times 100$

$= 2.3\%$ NaF

3. Use the same equation as for Problems 1 and 2, filling in given data:

$$\% \text{ Strength } = \frac{(\text{lbs Chem.}) \dfrac{(\% \text{ Comm. Purity})}{100}}{\text{lbs Water} + (\text{lbs Chem.}) \dfrac{(\% \text{ Comm. Pur.})}{100}} \times 100$$

$$1.5 = \frac{(x \text{ lbs}) \dfrac{(98)}{100}}{(200 \text{ gal}) (8.34 \text{ lbs/gal}) + (x \text{ lbs}) \dfrac{(98)}{100}} \times 100$$

$$1.5 = \frac{0.98x}{1668 + 0.98x} \times 100$$

$$1.5 = \frac{98x}{1668 + 0.98x}$$

$$1.5 (1668 + 0.98x) = 98x$$

$$2502 + 1.47x = 98x$$

$$2502 = 96.53x$$

$$25.9 \text{ lbs NaF} = x$$

4.

$$\% \text{ Strength } = \frac{(\text{lbs Chem.}) \dfrac{(\% \text{ Comm. Purity})}{100}}{\text{lbs Water} + (\text{lbs Chem.}) \dfrac{(\% \text{ Comm. Pur.})}{100}} \times 100$$

$$= \frac{(10 \text{ lbs}) \dfrac{(98)}{100}}{(50 \text{ gal}) (8.34 \text{ lbs/gal}) + (10 \text{ lbs}) \dfrac{(98)}{100}} \times 100$$

$$= \frac{9.8}{417 + 9.8} \times 100$$

$$= 2.3\% \text{ Strength NaF}$$

PRACTICE PROBLEMS—Cont'd

5. $\% \text{ Strength} = \dfrac{(\text{lbs Chem.}) \dfrac{(\% \text{ Comm. Purity})}{100}}{\text{lbs Water} + (\text{lbs Chem.}) \dfrac{(\% \text{ Comm. Pur.})}{100}} \times 100$

$2 = \dfrac{(x \text{ lbs}) \dfrac{(98)}{100}}{(150 \text{ gal}) (8.34 \text{ lbs/gal}) + (x \text{ lbs}) \dfrac{(98)}{100}} \times 100$

$2 = \dfrac{0.98x}{1251 + 0.98x} \times 100$

$2 = \dfrac{98x}{1251 + 0.98x}$

$2 (1251 + 0.98x) = 98x$

$2502 + 1.96x = 98x$

$2502 = 98x - 1.96x$

$2502 = 96.53x$

$26.1 \text{ lbs NaF} = x$

PRACTICE PROBLEMS 13.5

1. $\underset{\substack{\text{F} \quad \text{flow} \quad \text{lbs/gal}}}{(\text{mg}/L) \ (\text{MGD}) \ (8.34)} = \underset{\substack{\text{Comp.} \ \text{flow} \ \text{lbs/gal}}}{(\text{mg}/L) \ (\text{MGD}) \ (8.34)} \ (\text{sp. gr.}) \ \underset{\substack{\text{in Comp.}}}{\dfrac{(\% \text{ F})}{100}}$

$\underset{\substack{\text{mg}/L \ \text{MGD} \ \text{lbs/gal}}}{(1.1) \ (4.12) \ (8.34)} = \underset{\substack{\text{mg}/L \qquad\qquad \text{lbs/gal}}}{(220,000) \ (x \text{ MGD}) \ (8.34) \ (1.2)} \dfrac{(79.2)}{100}$

$\dfrac{(1.1) \ (4.12) \ \cancel{(8.34)}}{(220,000) \cancel{(8.34)} (1.2)(0.792)} = x \text{ MGD}$

$0.0000215 \text{ MGD} = x$

$21.5 \text{ gpd} = x$

2. $\underset{\substack{\text{F} \quad \text{flow} \quad \text{lbs/gal}}}{(\text{mg}/L) \ (\text{MGD}) \ (8.34)} = \underset{\substack{\text{Comp.} \ \text{flow} \ \text{lbs/gal}}}{(\text{mg}/L) \ (\text{MGD}) \ (\text{Sol'n})} \ \underset{\substack{\text{Comm.}}}{\dfrac{(\% \text{ F in})}{100}}$

$\underset{\substack{\text{mg}/L \ \text{MGD} \ \text{lbs/gal}}}{(1.1) \ (2.8) \ (8.34)} = \underset{\substack{\text{mg}/L \qquad\qquad \text{lbs/gal}}}{(200,000) \ (x \text{ MGD}) \ (9.8)} \dfrac{(79.2)}{100}$

$\dfrac{(1.1) \ (2.8) \ (8.34)}{(200,000) \ (9.8) \ (0.792)} = x \text{ MGD}$

$0.0000165 \text{ MGD} = x$

$16.5 \text{ gpd} = x$

3. The fluoride ion to be added to the water is 1.6 mg/L – 0.08 mg/L = 1.52 mg/L

$$
\begin{array}{cc}
\text{(mg/L) (MGD) (8.34)} & = \text{(mg/L) (MGD) (Sol'n) (% F in)} \\
\text{F \ \ flow \ \ lbs/gal} & \text{Comp. flow \ lbs/gal \ \underline{Comm.}} \\
& 100
\end{array}
$$

$$
\begin{array}{cc}
\text{(1.52) (0.84) (8.34)} & = \text{(21,000) (x MGD) (8.34) (45.25)} \\
\text{mg/L \ MGD \ lbs/gal} & \text{mg/L \ \ \ \ \ \ \ lbs/gal \ } \overline{100}
\end{array}
$$

$$
\frac{(1.52)\,(0.84)\,\cancel{(8.34)}}{(21,000)\,\cancel{(8.34)}\,(0.4525)} = x\,\text{MGD}
$$

$$
0.0001343\ \text{MGD} = x
$$

$$
134.3\ \text{gpd} = x
$$

4.
$$
\begin{array}{cc}
\text{(mg/L) (MGD) (8.34)} & = \text{(mg/L) (MGD) (Sol'n) (% F in)} \\
\text{F \ \ flow \ \ lbs/gal} & \text{Comp. flow \ lbs/gal \ \underline{Comm.}} \\
& 100
\end{array}
$$

$$
\begin{array}{cc}
\text{(1.5) (1.475) (8.34)} & = \text{(22,000) (x MGD) (8.34) (45.25)} \\
\text{mg/L \ MGD \ lbs/gal} & \text{mg/L \ \ \ \ \ \ \ lbs/gal \ } \overline{100}
\end{array}
$$

$$
\frac{(1.5)\,(1.475)\,\cancel{(8.34)}}{(22,000)\,\cancel{(8.34)}\,(0.4525)} = x\,\text{MGD}
$$

$$
0.0002222\ \text{MGD} = x
$$

$$
222.2\ \text{gpd} = x
$$

5. $\dfrac{\text{(90 gpd) (3785 mL/gal)}}{1440\ \text{min/day}} = 236.6\ \text{mL/min}$

6. First calculate the gpd solution feed rate:

$$
\begin{array}{cc}
\text{(1.0) (2.68) (8.34)} & = \text{(250,000) (x MGD) (9.8) (79.2)} \\
\text{mg/L \ MGD \ lbs/gal} & \text{mg/L \ \ \ \ \ \ \ lbs/gal \ } \overline{100}
\end{array}
$$

$$
\frac{(1.0)\,(2.68)\,(8.34)}{(250,000)\,(9.8)\,(0.792)} = x\,\text{MGD}
$$

$$
0.0000115\ \text{MGD} = x
$$

$$
11.5\ \text{gpd} = x
$$

Then convert gpd feed rate to mL/min feed rate:

$$
\frac{\text{(11.5 gpd) (3785 mL/gal)}}{1440\ \text{min/day}} = 30.2\ \text{mL/min}
$$

PRACTICE PROBLEMS 13.6

1. The approximate flow rate reading from the chart is 27 gpd.

    A proportion must be used to determine the flow rate corresponding to 1.3 mg/*L*:

    $$\frac{\text{Desired Dose, mg/}L}{\text{Chart Dose, 1 mg/}L} = \frac{\text{Desired Feed Rate, gpd}}{\text{Chart Feed Rate, gpd}}$$

    $$\frac{1.3 \text{ mg/}L}{1.0 \text{ mg/}L} = \frac{x \text{ gpd}}{27 \text{ gpd}}$$

    $$\frac{(1.3)(27)}{1} = x$$

    $$35.1 \text{ gpd} = x$$

2. The approximate flow rate reading from the chart is 100 gpd.

    A proportion must be used to determine the flow rate corresponding to 1.5 mg/*L* fluoride ion concentration:

    $$\frac{\text{Desired Dose, mg/}L}{\text{Chart Dose, 1 mg/}L} = \frac{\text{Desired Feed Rate, gpd}}{\text{Chart Feed Rate, gpd}}$$

    $$\frac{1.5 \text{ mg/}L}{1 \text{ mg/}L} = \frac{x \text{ gpd}}{100 \text{ gpd}}$$

    $$(1.5)(100) = x$$

    $$150 \text{ gpd} = x$$

3. The nomograph reading for a 0.7 MGD flow rate and fluoride level of 0.9 mg/*L* is approximately 36.5 gpd.

4. The fluoride ion to be added to the water is 1.1 mg/*L* – 0.07 mg/*L* = 1.03 mg/*L*. The approximate flow rate reading from the chart is 19 gpd.

    Use a proportion to determine the flow rate corresponding to 1.03 mg/*L*:

    $$\frac{\text{Desired Dose, mg/}L}{\text{Chart Dose, 1 mg/}L} = \frac{\text{Desired Feed Rate, gpd}}{\text{Chart Feed Rate, gpd}}$$

    $$\frac{1.03 \text{ mg/}L}{1 \text{ mg/}L} = \frac{x \text{ gpd}}{19 \text{ gpd}}$$

    $$(1.03)(19) = x$$

    $$19.6 \text{ gpd} = x$$

5. The fluoride ion to be added to the water is 1.6 mg/L – 0.1 mg/L = 1.5 mg/L. The approximate feed rate reading from the chart is 5 gph or (5 gph) (24 hrs/day) = 120 gpd.

    A proportion is used to determine the feed rate that corresponds to a fluoride level of 1.5 mg/L.

$$\frac{\text{Desired Dose, mg}/L}{\text{Chart Dose, 1 mg}/L} = \frac{\text{Desired Feed Rate, gpd}}{\text{Chart Feed Rate, gpd}}$$

$$\frac{1.5 \text{ mg}/L}{1 \text{ mg}/L} = \frac{x \text{ gpd}}{120 \text{ gpd}}$$

$$(1.5)(120) = x$$

$$180 \text{ gpd} = x$$

6. The nomograph reading for a flow rate of 400 gpm and a fluoride level of 1.1 mg/L is approximately 2.1 gpd hydrofluosilicic acid.

PRACTICE PROBLEMS 13.7

1. This is a calculation involving dry feed rate. Therefore the equation to be used is:

$$\frac{(\text{mg}/L \text{ F}) (\text{MGD flow}) (8.34 \text{ lbs/gal})}{\dfrac{(\% \text{ Comm. Pur.})}{100} \dfrac{(\% \text{ F in Comp.})}{100}} = \frac{\text{lbs/day}}{\text{Compound}}$$

Fill in the equation with the given data, then solve for the unknown value:

$$\frac{(x \text{ mg}/L) (1.48 \text{ MGD}) (8.34 \text{ lbs/gal})}{\dfrac{(98)}{100} \dfrac{(60.7)}{100}} = 30 \text{ lbs/day}$$

$$(x)(1.48)(8.34) = (30)(0.98)(0.607)$$

$$x = \frac{(30)(0.98)(0.607)}{(1.48)(8.34)}$$

$$x = 1.45 \text{ mg}/L \text{ F}$$

2. This is a dry feed rate calculation, so the following equation is used:

$$\frac{(\text{mg}/L \text{ F}) (\text{MGDflow}) (8.34 \text{ lbs/gal})}{\dfrac{(\% \text{ Comm. Pur.})}{100} \dfrac{(\% \text{ F in Comp.})}{100}} = \frac{\text{lbs/day}}{\text{Compound}}$$

Fill in the given information and solve for the unknown value:

$$\frac{(x \text{ mg}/L) (0.32 \text{ MGD}) (8.34 \text{ lbs/gal})}{\dfrac{(98)}{100} \dfrac{(45.25)}{100}} = 5 \text{ lbs/day}$$

$$(x)(0.32)(8.34) = (5)(0.98)(0.4525)$$

$$x = \frac{(5)(0.98)(0.4525)}{(0.32)(8.34)}$$

$$x = 0.83 \text{ mg}/L \text{ F}$$

PRACTICE PROBLEMS—Cont'd

3. Use the solution feed equation for this calculation:

$$\underset{\text{flow}}{(mg/L\ F)}\ (MGD)\ \underset{\text{lbs/gal}}{(8.34)} = \underset{\text{Comp. flow}}{(mg/L)}\ (MGD)\ \underset{\text{lbs/gal}}{(Sol'n)}\ \frac{(\%\ F\ \text{in Comp})}{100}$$

$$\underset{\text{MGD}}{(x\ mg/L)}\ (3.76)\ \underset{\text{lbs/gal}}{(8.34)} = \underset{\text{mg/L}}{(200,000)}\ (0.000025)\ \underset{\text{MGD}}{(9.8)}\ \underset{\text{lbs/gal}}{\frac{(79.2)}{100}}$$

$$x = \frac{(200,000)\ (0.000025)\ (9.8)\ (0.792)}{(3.76)\ (8.34)}$$

$$x = 1.2\ mg/L\ F$$

4. Use the solution feed equation for this calculation:

$$\underset{\text{flow}}{(mg/L\ F)}\ (MGD)\ \underset{\text{lbs/gal}}{(8.34)} = \underset{\text{Comp. flow}}{(mg/L)}\ (MGD)\ \underset{\text{lbs/gal}}{(Sol'n)}\ \frac{(\%\ F\ \text{in Comp})}{100}$$

$$\underset{\text{MGD}}{(x\ mg/L)}\ (1.84)\ \underset{\text{lbs/gal}}{(8.34)} = \underset{\text{mg/L}}{(100,000)}\ (0.000025)\ \underset{\text{MGD}}{(9.04)}\ \underset{\text{lbs/gal}}{\frac{(79.2)}{100}}$$

$$x = \frac{(100,000)\ (0.000025)\ (9.04)\ (0.792)}{(1.84)\ (8.34)}$$

$$x = 1.2\ mg/L\ F$$

5. First calculate the mg/L fluoride ion <u>added</u> to the water:

$$\underset{\text{flow}}{(mg/L\ F)}\ (MGD)\ \underset{\text{lbs/gal}}{(8.34)} = \underset{\text{Comp. flow}}{(mg/L)}\ (MGD)\ \underset{\text{lbs/gal}}{(Sol'n)}\ \frac{(\%\ F\ \text{in Comp})}{100}$$

$$\underset{\text{MGD}}{(x\ mg/L)}\ (2.62)\ \underset{\text{lbs/gal}}{(8.34)} = \underset{\text{mg/L}}{(40,000)}\ (0.000120)\ \underset{\text{MGD}}{(8.34)}\ \underset{\text{lbs/gal}}{\frac{(45.25)}{100}}$$

$$x = \frac{(40,000)\ (0.000120)\ (8.34)\ (0.4525)}{(2.62)\ (8.34)}$$

$$x = 0.83\ mg/L\ F$$

Now the total fluoride ion level in the finished water can be determined:

$$\underset{\text{Added}}{0.83\ mg/L\ F} + \underset{\text{in raw water}}{0.1\ mg/L\ F} = \underset{\text{finished water}}{0.93\ mg/L\ \text{in}}$$

PRACTICE PROBLEMS 13.8

1. % Strength of Mixture $= \dfrac{\dfrac{\text{lbs Chem. in}}{\text{Sol'n 1}} + \dfrac{\text{lbs Chem. in}}{\text{Sol'n 2}}}{\text{lbs Sol'n 1} + \text{lbs Sol'n 2}} \times 100$

$= \dfrac{(500\ \text{lbs})\ \dfrac{(15)}{100} + (2400\ \text{lbs})\ \dfrac{(25)}{100}}{500\ \text{lbs} + 2400\ \text{lbs}} \times 100$

$= \dfrac{75\ \text{lbs} + 600\ \text{lbs}}{2900\ \text{lbs}} \times 100$

$= 23.3\%$

2. In this calcualtion the alternative equation for solution mixtures will be used: (The equation used in Problem #1 may also be used for this calculation.)

$$R_1 C_1 + R_2 C_2 = R_3 C_3$$

$$(800\ \text{lbs})\dfrac{(25)}{100} + (250\ \text{lbs})\ \dfrac{(15)}{100} = (1050\ \text{lbs})\ \dfrac{(x)}{100}$$

$$200\ \text{lbs} + 37.5\ \text{lbs} = \dfrac{1050x}{100}$$

$$237.5 = 10.5x$$

$$\dfrac{237.5}{10.5} = x$$

$$22.6\% = x$$

3.

% Strength of Mixture $= \dfrac{\dfrac{\substack{(\text{Sol'n 1,})\ (\text{Dens.})\ (\%\ \text{Str.,})\\ \text{gal} \quad \text{lbs/gal} \quad \text{Sol'n 1}}}{100} + \dfrac{\substack{(\text{Sol'n 2,})\ (\text{Dens.})\ (\%\ \text{Str.,})\\ \text{gal} \quad \text{lbs/gal} \quad \text{Sol'n 2}}}{100}}{\substack{(\text{Sol'n 1,})\ (\text{Dens.,})\\ \text{gal} \quad \text{lbs/gal}} + \substack{(\text{Sol'n 2,})\ (\text{Dens.,})\\ \text{gal} \quad \text{lbs/gal}}} \times 100$

$= \dfrac{(350\ \text{gal})\ (9.6\ \text{lbs/gal})\ \dfrac{(18)}{100} + (2000)\ (9.8\ \text{lbs/gal})\ \dfrac{(20)}{100}}{(350\ \text{gal})\ (9.6\ \text{lbs/gal}) + (2000)\ (9.8\ \text{lbs/gal})} \times 100$

$= \dfrac{604.8\ \text{lbs} + 3920\ \text{lbs}}{3360\ \text{lbs} + 19{,}600\ \text{lbs}} \times 100$

$= \dfrac{4524.8\ \text{lbs}}{22{,}960\ \text{lbs}} \times 100$

$= 19.7\%$

PRACTICE PROBLEMS—Cont'd

4.

$$\% \text{ Strength of Mixture} = \frac{\dfrac{(\text{Sol'n 1,}) \ (\text{Dens.}) \ (\% \text{ Str.,})}{\text{gal} \quad \text{lbs/gal} \ \dfrac{\text{Sol'n 1}}{100}} + \dfrac{(\text{Sol'n 2,}) \ (\text{Dens.}) \ (\% \text{ Str.,})}{\text{gal} \quad \text{lbs/gal} \ \dfrac{\text{Sol'n 2}}{100}}}{\dfrac{(\text{Sol'n 1,}) \ (\text{Dens.,})}{\text{gal} \quad \text{lbs/gal}} + \dfrac{(\text{Sol'n 2,}) \ (\text{Dens.,})}{\text{gal} \quad \text{lbs/gal}}} \times 100$$

$$= \frac{(250 \text{ gal}) \ (9.04 \text{ lbs/gal}) \ \dfrac{(10)}{100} + (1000) \ (9.8 \text{ lbs/gal}) \ \dfrac{(20)}{100}}{(250 \text{ gal}) \ (9.04 \text{ lbs/gal}) + (1000) \ (9.8 \text{ lbs/gal})} \times 100$$

$$= \frac{226 \text{ lbs} + 1960 \text{ lbs}}{22604 \text{ lbs} + 9800 \text{ lbs}} \times 100$$

$$= \frac{2186 \text{ lbs}}{12,060 \text{ lbs}} \times 100$$

$$= 18.1\%$$

5. The density of the 200 gallon solution can be determined: (8.34 lbs/gal)(1.075) = 8.97 lbs/gal. Then continue as usual:

$$\% \text{ Strength of Mixture} = \frac{\dfrac{(\text{Sol'n 1,}) \ (\text{Dens.}) \ (\% \text{ Str.,})}{\text{gal} \quad \text{lbs/gal} \ \dfrac{\text{Sol'n 1}}{100}} + \dfrac{(\text{Sol'n 2,}) \ (\text{Dens.}) \ (\% \text{ Str.,})}{\text{gal} \quad \text{lbs/gal} \ \dfrac{\text{Sol'n 2}}{100}}}{\dfrac{(\text{Sol'n 1,}) \ (\text{Dens.,})}{\text{gal} \quad \text{lbs/gal}} + \dfrac{(\text{Sol'n 2,}) \ (\text{Dens.,})}{\text{gal} \quad \text{lbs/gal}}} \times 100$$

$$= \frac{(200 \text{ gal}) \ (8.97 \text{ lbs/gal}) \ \dfrac{(9)}{100} + (1500) \ (9.4 \text{ lbs/gal}) \ \dfrac{(15)}{100}}{(200 \text{ gal}) \ (8.97 \text{ lbs/gal}) + (1500) \ (9.4 \text{ lbs/gal})} \times 100$$

$$= \frac{161.5 \text{ lbs} + 2115 \text{ lbs}}{1794 \text{ lbs} + 14,100 \text{ lbs}} \times 100$$

$$= \frac{2276.5 \text{ lbs}}{15,894 \text{ lbs}} \times 100$$

$$= 14.3\%$$

CHAPTER 13 ACHIEVEMENT TEST

1. $2.8\% = 2.8000. = 28{,}000$ mg/L

   $= 28{,}000$ mg/L

2. First calculate the molecular weight of each element in the compound:

Element	No. of Atoms	Atomic Wt.	Molec. Wt.
H:	2	x 1.008 =	2.016
Si:	1	x 28.06 =	28.06
F:	6	x 19.00 =	114.00
			144.076

   The percent fluoride in the compound can now be calculated:

   $$\text{\% F in } H_2SiF_6 = \frac{\text{Molec. Wt. of F}}{\text{Molec. Wt. of } H_2SiF_6} \times 100$$

   $$= \frac{114.00}{144.076} \times 100$$

   $$= 79.1\%$$

3. $\dfrac{24 \text{ lbs}}{(1 \text{ MG}) (8.34 \text{ lbs/gal})} = \dfrac{24 \text{ lbs}}{8.34 \text{ mil. lbs}} = \dfrac{2.9 \text{ lbs}}{1 \text{ mil. lbs}} = 2.9$ mg/L

4. $\dfrac{(1.4 \text{ mg/}L) (1.156 \text{ MGD}) (8.34 \text{ lbs/gal})}{\dfrac{(98)}{100} \dfrac{(60.6)}{100}} = \dfrac{22.7 \text{ lbs/day}}{Na_2SiF_6}$

5. First calculate the molecular weight of each element in the compound:

Element	No. of Atoms	Atomic Wt.	Molec. Wt.
Na:	1	22.997	22.997
F:	1	19.00	19.00
		Molec. Wt. of NaF:	41.997

   Then calculate the percent fluoride ion in the compound:

   $$\text{\% F in NaF} = \frac{\text{Molec. Wt. of F}}{\text{Molec. Wt. of NaF}} \times 100$$

   $$= \frac{19.00}{41.997} \times 100$$

   $$= 45.2\%$$

ACHIEVEMENT TEST—Cont'd

6. $\% \text{ Strength} = \dfrac{\dfrac{(\text{lbs Chem.}) \ (\% \text{ Comm. Purity})}{100}}{\text{lbs Water} + \dfrac{(\text{lbs Chem.}) \ (\% \text{ Comm. Pur.})}{100}} \times 100$

$= \dfrac{\dfrac{(70 \text{ lbs}) \ (98)}{100}}{(500 \text{ gal}) \ (8.34 \text{ lbs/gal}) + \dfrac{(70 \text{ lbs}) \ (98)}{100}} \times 100$

$= \dfrac{68.6 \text{ lbs}}{4170 \text{ lbs} + 68.6 \text{ lbs}} \times 100$

$= \dfrac{68.6}{4238.6} \times 100$

$= 1.6\%$

7. $25{,}000 \text{ mg}/L = 25{,}000.\%$

$= 2.5 \%$

8. $\dfrac{(75 \text{ gpd}) \ (3785 \text{ m}L/\text{gal})}{1440 \text{ min/day}} = 197 \text{ m}L/\text{min}$

9. $\dfrac{(1.3 \text{ mg}/L) \ (2.17 \text{ MGD}) \ (8.34 \text{ lbs/gal})}{\dfrac{(98)}{100} \ \dfrac{(45.25)}{100}} = 53 \text{ lbs/day}$

10. $\% \text{ Strength} = \dfrac{\dfrac{(\text{lbs Chem.}) \ (\% \text{ Comm. Purity})}{100}}{\text{lbs Water} + \dfrac{(\text{lbs Chem.}) \ (\% \text{ Comm. Pur.})}{100}} \times 100$

$2 = \dfrac{\dfrac{(x \text{ lbs}) \ (98)}{100}}{(400 \text{ gal}) \ (8.34 \text{ lbs/gal}) + \dfrac{(x \text{ lbs}) \ (98)}{100}} \times 100$

$2 = \dfrac{0.98x}{3336 \text{ lbs} + 0.98x} \times 100$

$2 = \dfrac{98x}{3336 + 0.98x}$

$2 \ (3336 + 0.98x) = 98x$

$6672 + 1.96x = 98x$

$6672 = 96.04x$

$69.5 \text{ lbs NaF} = x$

11. $\dfrac{(mg/L)}{F} \ \dfrac{(MGD)}{flow} \ \dfrac{(8.34)}{lbs/gal} = \dfrac{(mg/L)}{Comp.} \ \dfrac{(MGD)}{flow} \ \dfrac{(8.34)}{lbs/gal} \ (sp. \ gr.) \ \dfrac{(\% \ F)}{\dfrac{in \ Comp.}{100}}$

   $\dfrac{(1.5)}{mg/L} \ \dfrac{(3.96)}{MGD} \ \dfrac{(8.34)}{lbs/gal} = (220,000) \ (x \ MGD) \ \dfrac{(8.34)}{lbs/gal} \ (1.2) \ \dfrac{(79.2)}{100}$

   $\dfrac{(1.5) \ (3.96) \ \cancel{(8.34)}}{(220,000)\cancel{(8.34)}(1.2)(0.792)} = x \ MGD$

   $0.0000284 \ MGD = x$

   $28.4 \ gpd = x$

12. $\% \ Strength = \dfrac{(lbs \ Chem.) \ \dfrac{(\% \ Comm. \ Purity)}{100}}{lbs \ Water + (lbs \ Chem.) \ \dfrac{(\% \ Comm. \ Pur.)}{100}} \ x \ 100$

   $= \dfrac{(25 \ lbs) \ \dfrac{(98)}{100}}{(120 \ gal) \ (8.34 \ lbs/gal) + (25 \ lbs) \ \dfrac{(98)}{100}} \ x \ 100$

   $= \dfrac{24.5 \ lbs}{1000.8 \ lbs + 24.5 \ lbs} \ x \ 100$

   $= \dfrac{24.5}{1025.3} \ x \ 100$

   $= 2.4\%$

13. The flow rate reading from the chart is about 35 gpd.

   A proportion must be used to determine the flow rate corresponding to 1.4 mg/L:

   $\dfrac{Desired \ Dose, \ mg/L}{Chart \ Dose, \ 1 \ mg/L} = \dfrac{Desired \ Feed \ Rate, \ gpd}{Chart \ Feed \ Rate, \ gpd}$

   $\dfrac{1.4 \ mg/L}{1.0 \ mg/L} = \dfrac{x \ gpd}{35 \ gpd}$

   $(1.4) \ (35) = x$

   $49 \ gpd = x$

14. The fluoride ion to be added to the water is 1.3 mg/L – 0.08 mg/L = 1.22 mg/L.

   $\dfrac{(1.22 \ mg/L) \ (1.772 \ MGD) \ (8.34 \ lbs/gal)}{0.44} = \dfrac{41 \ lbs/day}{NaF}$

ACHIEVEMENT TEST—Cont'd

15. $$\underset{\substack{\text{F} \quad\text{flow} \quad\text{lbs/gal}}}{(mg/L)\ (MGD)\ (8.34)} = \underset{\substack{\text{Comp. flow} \quad\text{lbs/gal}}}{(mg/L)\ (MGD)\ (8.34)\ (\text{sp. gr.})} \underset{100}{\underbrace{(\% \text{ F})}_{\text{in Comp.}}}$$

$$\underset{\substack{mg/L \quad MGD \quad\text{lbs/gal}}}{(1.2)\ (2.6)\ (8.34)} = \underset{\substack{mg/L \quad\quad\quad\text{lbs/gal}}}{(200,000)\ (x\ MGD)\ (9.8)} \ \frac{(79.2)}{100}$$

$$\frac{(1.2)\ (2.6)\ (8.34)}{(200,000)(9.8)(0.792)} = x\ MGD$$

$$0.0000168\ MGD = x$$

$$16.8\ gpd = x$$

16. $$\underset{\substack{\text{\% Strength} \\ \text{of Mixture}}}{} = \frac{\dfrac{\text{lbs Chem. in}}{\text{Sol'n 1}} + \dfrac{\text{lbs Chem. in}}{\text{Sol'n 2}}}{\text{lbs Sol'n 1} + \text{lbs Sol'n 2}} \times 100$$

$$= \frac{(400\ \text{lbs})\ \dfrac{(15)}{100} + (1500\ \text{lbs})\ \dfrac{(20)}{100}}{400\ \text{lbs} + 1500\ \text{lbs}} \times 100$$

$$= \frac{60\ \text{lbs} + 300\ \text{lbs}}{1900\ \text{lbs}} \times 100$$

$$= \frac{360\ \text{lbs}}{1900\ \text{lbs}} \times 100$$

$$= 18.9\%$$

17. $$\frac{(mg/L\ \text{F})\ (MGD)\ (8.34\ \text{lbs/gal})}{\dfrac{(\% \text{ Comm. Pur.})}{100}\ \dfrac{(\% \text{ F in Comp.})}{100}} = \frac{\text{lbs/day}}{\text{Compound}}$$

Fill in the equation with the given data, then solve for the unknown value:

$$\frac{(x\ mg/L)\ (1.07\ MGD)\ (8.34\ \text{lbs/gal})}{\dfrac{(98)}{100}\ \dfrac{(60.7)}{100}} = 38\ \text{lbs/fay}$$

$$(x)\ (1.07)\ (8.34) = (38)\ (0.98)\ (0.607)$$

$$x = \frac{(38)\ (0.98)\ (0.607)}{(1.07)\ (8.34)}$$

$$x = 2.5\ mg/L\ \text{F}$$

18. Before beginning the calculation, convert 1300 gpm flow rate to MGD:

$$(1300 \text{ gpm}) (1440 \text{ min/day}) = 1,872,000 \text{ gpd}$$

$$= 1.872 \text{ MGD}$$

Then continue with the solution feed rate calculation:

$$
\begin{array}{l}
\underset{\text{F}}{(\text{mg}/L)} \ \underset{\text{flow}}{(\text{MGD})} \ \underset{\text{lbs/gal}}{(8.34)} = \underset{\text{Comp. flow}}{(\text{mg}/L)} \ (\text{MGD}) \ \underset{\text{lbs/gal}}{(\text{Sol'n})} \ \underset{\underline{\text{Comm.}}}{\underset{100}{(\% \text{ F in})}}
\end{array}
$$

$$
\underset{\text{mg}/L}{(x)} \ \underset{\text{flow}}{(1.872)} \ \underset{\text{lbs/gal}}{(8.34)} = \underset{\text{mg}/L}{(100,000)} \ (0000.30 \text{ MGD}) \ \underset{\text{lbs/gal}}{(9.04)} \ \frac{(79.2)}{100}
$$

$$x = \frac{(100,000) \ (0.000030) \ (9.04) \ (0.792)}{(1.872) \ (8.34)}$$

$$x = 1.38 \text{ mg}/L$$

19. The nomograph reading for a 300 gpm flow rate and fluoride level of 1.1 mg/L is about 27 gpd NaF.

20. % Strength of Mixture =

$$
\frac{\underset{\text{gal}}{(\text{Sol'n 1,})} \ \underset{\text{lbs/gal}}{(\text{Dens.})} \ \frac{(\% \text{ Str.,})}{\underset{100}{\text{Sol'n 1}}} + \underset{\text{gal}}{(\text{Sol'n 2,})} \ \underset{\text{lbs/gal}}{(\text{Dens.})} \ \frac{(\% \text{ Str.,})}{\underset{100}{\text{Sol'n 2}}}}{\underset{\text{gal}}{(\text{Sol'n 1,})} \ \underset{\text{lbs/gal}}{(\text{Dens.,})} + \underset{\text{gal}}{(\text{Sol'n 2,})} \ \underset{\text{lbs/gal}}{(\text{Dens.,})}} \times 100
$$

$$
= \frac{(215 \text{ gal}) (9.04 \text{ lbs/gal}) \frac{(10)}{100} + (500 \text{ gal}) (9.8 \text{ lbs/gal}) \frac{(20)}{100}}{(215 \text{ gal}) (9.04 \text{ lbs/gal}) + (500 \text{ gal}) (9.8 \text{ lbs/gal})} \times 100
$$

$$
= \frac{194.4 \text{ lbs} + 980 \text{ lbs}}{1943.6 \text{ lbs} + 4900 \text{ lbs}} \times 100
$$

$$
= \frac{1174.4 \text{ lbs}}{6843.6 \text{ lbs}} \times 100
$$

$$
= 17.2\%
$$

ACHIEVEMENT TEST—Cont'd

21. First calculate the gpd solution feed rate:

$$\underset{\substack{\text{mg/}L \quad \text{MGD} \quad \text{lbs/gal}}}{(1.1)\ (2.08)\ (8.34)} = \underset{\substack{\text{mg/}L \quad\quad\quad \text{lbs/gal}}}{(250{,}000)\ (x\,\text{MGD})\ (\ 9.8\ )\ \dfrac{(79.2)}{100}}$$

$$\frac{(1.1)\ (2.08)\ (8.34)}{(250{,}000)(9.8)(0.792)} = x\,\text{MGD}$$

$$0.00000983\ \text{MGD} = x$$

$$9.8\ \text{gpd} = x$$

Then convert gpd feed rate to mL/min feed rate:

$$\frac{(9.8\ \text{gpd})\ (3785\ \text{m}L/\text{gal})}{1440\ \text{min/day}} = 25.8\ \text{m}L/\text{min}$$

22. The density of the 128-gallon solution can be determined:

$$(8.34\ \text{lbs/gal})\ (1.075) = 8.97\ \text{lbs/gal}$$

Then continue with the percent strength calculation:

$$\begin{array}{l}\text{\% Strength}\\ \text{of Mixture}\end{array} = \frac{\dfrac{\underset{\substack{\text{gal} \quad\ \text{lbs/gal} \quad \text{Sol'n 1}}}{(\text{Sol'n 1,})\ (\text{Dens.})\ (\text{\% Str.,})}}{100} + \dfrac{\underset{\substack{\text{gal} \quad\ \text{lbs/gal} \quad \text{Sol'n 2}}}{(\text{Sol'n 2,})\ (\text{Dens.})\ (\text{\% Str.,})}}{100}}{\underset{\substack{\text{gal} \quad \text{lbs/gal}}}{(\text{Sol'n 1,})\ (\text{Dens.,})} + \underset{\substack{\text{gal} \quad \text{lbs/gal}}}{(\text{Sol'n 2,})\ (\text{Dens.,})}} \times 100$$

$$= \frac{(128\ \text{gal})\ (8.97\ \text{lbs/gal})\ \dfrac{(\ 9\ )}{100} + (800\ \text{gal})\ (9.4\ \text{lbs/gal})\ \dfrac{(15)}{100}}{(128\ \text{gal})\ (8.97\ \text{lbs/gal}) + (800\ \text{gal})\ (9.4\ \text{lbs/gal})} \times 100$$

$$= \frac{103.3\ \text{lbs} + 1128\ \text{lbs}}{1148.2\ \text{lbs} + 7520\ \text{lbs}} \times 100$$

$$= \frac{1231.3\ \text{lbs}}{8668.2\ \text{lbs}} \times 100$$

$$= 14.2\%$$

23. The fluoride ion to be added to the water is 1.4 mg/L – 0.1 mg/L = 1.3 mg/L.
    The approximate feed rate reading from the chart is 152 gpd. A proportion is used to determine the feed rate that corresponds to a fluoride level of 1.3 mg/L:

$$\frac{1.3\ \text{mg/}L}{1.0\ \text{mg/}L} = \frac{x\ \text{gpd}}{152\ \text{gpd}}$$

$$(1.3)\ (152) = x$$

$$197.6\ \text{gpd} = x$$

24.   First calculate the mg/*L* fluoride <u>added</u> to the water:

(mg/*L*) (MGD flow) (8.34 lbs/gal) = (mg/*L*) (MGD flow) (Sol'n, lbs/gal) $\dfrac{(\% \text{ F in Comp.})}{100}$

(*x* mg/*L*) (2.88 MGD flow) (8.34 lbs/gal) = (40,000 mg/*L*) (0.000125 MGD) (8.34 lbs/gal) $\dfrac{(45.25)}{100}$

$$x = \frac{(40,000)\,(0.000125)\,\cancel{(8.34)}\,(0.4525)}{(2.88)\,\cancel{(8.34)}}$$

$$x = 0.79 \text{ mg/}L \text{ F}$$

Now calculate the total fluoride ion level in the finished water:

$$\underset{\text{Added}}{0.79 \text{ mg/}L \text{ F}} + \underset{\text{in raw water}}{0.1 \text{ mg/}L} = \underset{\text{finished water}}{0.89 \text{ mg/}L \text{ in}}$$

Chapter 14

PRACTICE PROBLEMS—14.1

1.   Equivalent Weight of Na $= \dfrac{\text{Atomic Weight}}{\text{Valence}}$

$$= \frac{22.997}{1}$$

$$= 22.997$$

2.   Equivalent Weight of Ca $= \dfrac{\text{Atomic Weight}}{\text{Valence}}$

$$= \frac{40.08}{2}$$

$$= 20.04$$

3.   Equivalent Weight of Mg $= \dfrac{\text{Atomic Weight}}{\text{Valence}}$

$$= \frac{24.32}{2}$$

$$= 12.16$$

4.   First calculate the molecular weight of NaCl:

Na: 22.997 x 1 = 22.997

Cl: 35.457 x 1 = $\underline{35.457}$
              58.454 Molecular Weight

Then calculate the equivalent weight of NaCl: (The valence of Na is 1.)

Equivalent Weight of NaCl $= \dfrac{\text{Molecular Weight}}{\text{Net Positive Valence}}$

$$= \frac{58.454}{1}$$

$$= 58.454$$

PRACTICE PROBLEMS—Cont'd

5. Calculate the molecular weight of $CaCO_3$:

$$
\begin{aligned}
Ca &= 40.08 \quad x\,1 = 40.08 \\
C &= 12.010 \quad x\,1 = 12.010 \\
O_3 &= 16.000 \quad x\,3 = \underline{48.000} \\
&\qquad\qquad\qquad\quad 100.090
\end{aligned}
$$

Then calculate the equivalent weight of $CaCO_3$: (The valence of Ca is +2)

$$
\text{Equivalent Weight of } CaCO_3 = \frac{\text{Molecular Weight}}{\text{Net Positive Valence}}
$$

$$
= \frac{100.090}{2}
$$

$$
= 50.045
$$

6. $\text{Number of Milliequivalents} = \dfrac{\text{Chemical, mg}}{\text{Equivalent Wt.}}$

$$
= \frac{275 \text{ mg}}{20.04}
$$

$$
= 13.7 \text{ milliequivalents}
$$

7. $\text{Number of Milliequivalents per liter} = \dfrac{\text{Chemical, mg/}L}{\text{Equivalent Wt.}}$

$$
= \frac{32 \text{ mg/}L}{12.16}
$$

$$
= 2.6 \text{ milliequivalents/}L
$$

8. $\text{Number of Milliequivalents per liter} = \dfrac{\text{Chemical, mg/}L}{\text{Equivalent Wt.}}$

$$
= \frac{30 \text{ mg/}L}{12.15}
$$

$$
= 2.5 \text{ milliequivalents/}L
$$

PRACTICE PROBLEMS 14.2

1. $$\frac{\text{Calcium Hardness,}\\ \text{mg/}L \text{ as CaCO}_3}{\text{Equivalent Wt. of CaCO}_3} = \frac{\text{Calcium, mg/}L}{\text{Equivalent Weight of Calcium}}$$

    $$\frac{x \text{ mg/}L}{50.045} = \frac{37 \text{ mg/}L}{20.04}$$

    Then solve for the unknown value:

    $$x = \frac{(37)(50.045)}{20.04}$$

    $$x = 92.4 \text{ mg/}L \text{ Calcium}\\ \text{as CaCO}_3$$

2. $$\frac{\text{Magnesium Hardness,}\\ \text{mg/}L \text{ as CaCO}_3}{\text{Equivalent Wt. of CaCO}_3} = \frac{\text{Magnesium, mg/}L}{\text{Equivalent Wt. of Magnesium}}$$

    $$\frac{x \text{ mg/}L}{50.045} = \frac{31 \text{ mg/}L}{12.16}$$

    $$x = \frac{(31)(50.045)}{12.16}$$

    $$x = 127.6 \text{ mg/}L \text{ Magnesium}\\ \text{as CaCO}_3$$

3. $$\frac{\text{Calcium Hardness,}\\ \text{mg/}L \text{ as CaCO}_3}{\text{Equivalent Wt. of CaCO}_3} = \frac{\text{Calcium, mg/}L}{\text{Equivalent Weight of Calcium}}$$

    $$\frac{x \text{ mg/}L}{50.045} = \frac{19 \text{ mg/}L}{20.04}$$

    Then solve for the unknown value:

    $$x = \frac{(19)(50.045)}{20.04}$$

    $$x = 47.4 \text{ mg/}L \text{ Calcium}\\ \text{as CaCO}_3$$

4. $$\begin{array}{c}\text{Total Hardness,}\\ \text{mg/}L \text{ as CaCO}_3\end{array} = \begin{array}{c}\text{Calcium Hardness,}\\ \text{mg/}L \text{ as CaCO}_3\end{array} + \begin{array}{c}\text{Magnesium Hardness,}\\ \text{mg/}L \text{ as CaCO}_3\end{array}$$

    $$= 72 \text{ mg/}L + 89 \text{ mg/}L$$

    $$= 161 \text{ mg/}L \text{ as CaCO}_3$$

PRACTICE PROBLEMS—Cont'd

5. First express the calcium and magnesium hardness in mg/L as $CaCO_3$:

$$\frac{\text{Calcium Hardness,}}{\text{Equivalent Wt. of CaCO}_3} = \frac{\text{Calcium, mg/L}}{\text{Equivalent Weight of Calcium}}$$

$$\frac{x \text{ mg/L}}{50.045} = \frac{25 \text{ mg/L}}{20.04}$$

Then solve for the unknown value:

$$x = \frac{(25)(50.045)}{20.04}$$

$$x = 62.4 \text{ mg/L Calcium} \atop \text{as CaCO}_3$$

$$\frac{\text{Magnesium Hardness,}}{\text{Equivalent Wt. of CaCO}_3} = \frac{\text{Magnesium, mg/L}}{\text{Equivalent Wt of Magnesium}}$$

$$\frac{x \text{ mg/L}}{50.045} = \frac{9 \text{ mg/L}}{12.16}$$

$$x = \frac{(9)(50.045)}{12.16}$$

$$x = 37 \text{ mg/L Magnesium} \atop \text{as CaCO}_3$$

Then calculate the total hardness as $CaCO_3$:

$$\frac{\text{Total Hardness,}}{\text{mg/L as CaCO}_3} = \frac{\text{Calcium Hardness,}}{\text{mg/L as CaCO}_3} + \frac{\text{Magnesium Hardness,}}{\text{mg/L as CaCO}_3}$$

$$= 62.4 \text{ mg/L} + 37 \text{ mg/L}$$

$$= 99.4 \text{ mg/L total hardness}$$

6. First express calcium and magnesium hardness in mg/L as $CaCO_3$:

$$\frac{\text{Calcium Hardness,}\ \text{mg/}L\ \text{as}\ CaCO_3}{\text{Equivalent Wt. of}\ CaCO_3} = \frac{\text{Calcium, mg/}L}{\text{Equivalent Weight of Calcium}}$$

$$\frac{x\ \text{mg/}L}{50.045} = \frac{19\ \text{mg/}L}{20.04}$$

Then solve for the unknown value:

$$x = \frac{(19)\ (50.045)}{20.04}$$

$$x = 47.4\ \text{mg/}L\ \text{Calcium}$$
$$\text{as}\ CaCO_3$$

$$\frac{\text{Magnesium Hardness,}\ \text{mg/}L\ \text{as}\ CaCO_3}{\text{Equivalent Wt. of}\ CaCO_3} = \frac{\text{Magnesium, mg/}L}{\text{Equivalent Wt of Magnesium}}$$

$$\frac{x\ \text{mg/}L}{50.045} = \frac{14\ \text{mg/}L}{12.16}$$

$$x = \frac{(14)\ (50.045)}{12.16}$$

$$x = 57.6\ \text{mg/}L\ \text{Magnesium}$$
$$\text{as}\ CaCO_3$$

Now the total hardness can be calculated:

$$\begin{array}{l}\text{Total Hardness,}\\ \text{mg/}L\ \text{as}\ CaCO_3\end{array} = \begin{array}{l}\text{Calcium Hardness,}\\ \text{mg/}L\ \text{as}\ CaCO_3\end{array} + \begin{array}{l}\text{Magnesium Hardness,}\\ \text{mg/}L\ \text{as}\ CaCO_3\end{array}$$

$$= 47.4\ \text{mg/}L + 57.6\ \text{mg/}L$$

$$= 105\ \text{mg/}L\ \text{total hardness}$$
$$\text{as}\ CaCO_3$$

PRACTICE PROBLEMS 14.3

1. The alkalinity (120 mg/L as $CaCO_3$) is greater than the total hardness (105 mg/L as $CaCO_3$). Therefore, all the hardness is carbonate hardness:

$$\begin{array}{l}\text{Total Hardness,}\\ \text{mg/}L\ \text{as}\ CaCO_3\end{array} = \begin{array}{l}\text{Carbonate Hardness,}\\ \text{mg/}L\ \text{as}\ CaCO_3\end{array}$$

$$\begin{array}{l}105\ \text{mg/}L\\ \text{as}\ CaCO_3\end{array} = \begin{array}{l}\text{Carbonate}\\ \text{Hardness}\end{array}$$

There is no noncarbonate hardness in this water.

PRACTICE PROBLEMS—Cont'd

2. Alkalinity is less than total hardness. Therefore, both carbonate and noncarbonate water will be present in the water. The alkalinity content of the water represents the <u>carbonate hardness</u> of the water.

   Since both carbonate and total hardness data are known, the noncarbonate hardness can be calculated:

   $$\text{Total Hardness,} \atop \text{mg}/L \text{ as CaCO}_3 = \text{Carbonate Hardness,} \atop \text{mg}/L \text{ as CaCO}_3 + \text{Noncarbonate Hardness,} \atop \text{mg}/L \text{ as CaCO}_3$$

   $$118 \text{ mg}/L = 95 \text{ mg}/L + x \text{ mg}/L$$

   $$118 \text{ mg}/L - 95 \text{ mg}/L = x \text{ mg}/L$$

   $$23 \text{ mg}/L = x \atop \text{noncarbonate hardness}$$

   (Carbonate hardness is 95 mg/L, as indicated above.)

3. $$\text{Total Hardness,} \atop \text{mg}/L \text{ as CaCO}_3 = \text{Carbonate Hardness,} \atop \text{mg}/L \text{ as CaCO}_3 + \text{Noncarbonate Hardness,} \atop \text{mg}/L \text{ as CaCO}_3$$

   $$114 \text{ mg}/L = 82 \text{ mg}/L + x \text{ mg}/L$$

   $$114 \text{ mg}/L - 82 \text{ mg}/L = x \text{ mg}/L$$

   $$32 \text{ mg}/L = x \atop \text{noncarbonate hardness}$$

   (Carbonate hardness is 82 mg/L, as indicated above.)

4. In this problem alkalinity is greater than total hardness. Therefore, all the hardness is carbonate hardness:

   $$\text{Total Hardness,} \atop \text{mg}/L \text{ as CaCO}_3 = \text{Carbonate Hardness,} \atop \text{mg}/L \text{ as CaCO}_3$$

   $$97 \text{ mg}/L \atop \text{as CaCO}_3 = \text{Carbonate Hardness}$$

5. $$\text{Total Hardness,} \atop \text{mg}/L \text{ as CaCO}_3 = \text{Carbonate Hardness,} \atop \text{mg}/L \text{ as CaCO}_3 + \text{Noncarbonate Hardness,} \atop \text{mg}/L \text{ as CaCO}_3$$

   $$119 \text{ mg}/L = 101 \text{ mg}/L + x \text{ mg}/L$$

   $$119 \text{ mg}/L - 101 \text{ mg}/L = x \text{ mg}/L$$

   $$18 \text{ mg}/L = x \atop \text{noncarbonate hardness}$$

   (Carbonate hardness is 101 mg/L, as indicated above.)

PRACTICE PROBLEMS 14.4

1. Phenolphthalein Alkalinity mg/L as $CaCO_3$

$$= \frac{(A)(N)(50,000)}{mL \text{ of Sample}}$$

$$= \frac{(1.0 \text{ mL}) (0.02N) (50,000)}{100 \text{ mL}}$$

$$= 10 \text{ mg/L as } CaCO_3 \text{ Phenolphthalein Alkalinity}$$

2. Phenolphthalein Alkalinity mg/L as $CaCO_3$

$$= \frac{(A)(N)(50,000)}{mL \text{ of Sample}}$$

$$= \frac{(1.3 \text{ mL}) (0.02N) (50,000)}{100 \text{ mL}}$$

$$= 13 \text{ mg/L as } CaCO_3 \text{ Phenolphthalein Alkalinity}$$

3. Phenolphthalein Alkalinity mg/L as $CaCO_3$

$$= \frac{(0.2 \text{ mL}) (0.02N) (50,000)}{100 \text{ mL}}$$

$$= 2 \text{ mg/L}$$

Total Alkalinity mg/L as $CaCO_3$

$$= \frac{(6.9 \text{ mL}) (0.02N) (50,000)}{100 \text{ mL}}$$

$$= 69 \text{ mg/L}$$

4. Phenolphthalein Alkalinity mg/L as $CaCO_3$

$$= 0 \text{ mg/L}$$

Total Alkalinity mg/L as $CaCO_3$

$$= \frac{(7.1 \text{ mL}) (0.02N) (50,000)}{100 \text{ mL}}$$

$$= 71 \text{ mg/L}$$

5. Phenolphthalein Alkalinity mg/L as $CaCO_3$

$$= \frac{(0.4 \text{ mL}) (0.02N) (50,000)}{100 \text{ mL}}$$

$$= 4 \text{ mg/L}$$

Total Alkalinity mg/L as $CaCO_3$

$$= \frac{(5.9 \text{ mL}) (0.02N) (50,000)}{100 \text{ mL}}$$

$$= 59 \text{ mg/L}$$

PRACTICE PROBLEMS 14.5

1. P alkalinity (9 mg/L) is less than half of the T alkalinity (48 mg/L ÷ 2 = 24 mg/L)

   Bicarbonate Alkalinity = T – 2P

   $$= 48 \text{ mg}/L - 2 \ (9 \text{ mg}/L)$$

   $$= 48 \text{ mg}/L - 18 \text{ mg}/L$$

   $$= 30 \text{ mg}/L \text{ as } CaCO_3$$

   Carbonate Alkalinity = 2P

   $$= 2 \ (9 \text{ mg}/L)$$

   $$= 18 \text{ mg}/L \text{ as } CaCO_3$$

   Hydroxide Alkalinity = 0 mg/L

2. Bicarbonate Alkalinity = T

   $$= 68 \text{ mg}/L \text{ as } CaCO_3$$

   Carbonate Alkalinity = 0 mg/L

   Hydroxide Alkalinity = 0 mg/L

3. P alkalinity (14 mg/L) is greater than half of the T alkalinity (22 mg/L ÷ 2 = 11 mg/L)

   Bicarbonate Alkalinity = 0 mg/L

   Carbonate Alkalinity = 2T – 2P

   $$= 2 \ (22 \text{ mg}/L) - 2 \ (14 \text{ mg}/L)$$

   $$= 44 \text{ mg}/L - 28 \text{ mg}/L$$

   $$= 16 \text{ mg}/L \text{ as } CaCO_3$$

   Hydroxide Alkalinity = 2P – T

   $$= 2 \ (14 \text{ mg}/L) - 22 \text{ mg}/L$$

   $$= 28 \text{ mg}/L - 22 \text{ mg}/L$$

   $$= 6 \text{ mg}/L \text{ as } CaCO_3$$

4. Phenolphthalein Alkalinity
    mg/L as $CaCO_3$
    $$= \frac{(1.2 \text{ m}L)\,(0.02N)\,(50{,}000)}{100 \text{ m}L}$$

    $= 12$ mg/L as $CaCO_3$

    Total Alkalinity
    mg/L as $CaCO_3$
    $$= \frac{(5.1 \text{ m}L)\,(0.02N)\,(50{,}000)}{100 \text{ m}L}$$

    $= 51$ mg/L as $CaCO_3$

P is less than half T alkalinity. Use the alkalinity table to determine the remaining alkalinity constituents:

    Bicarbonate Alkalinity $= T - 2P$

    $= 51$ mg/$L - 2\,(12$ mg/$L)$

    $= 51$ mg/$L - 24$ mg/L

    $= 27$ mg/L as $CaCO_3$

    Carbonate Alkalinity $= 2P$

    $= 2\,(12$ mg/$L)$

    $= 24$ mg/L as $CaCO_3$

Hydroxide Alkalinity $= 0$ mg/L

5. Phenolphthalein Alkalinity
    mg/L as $CaCO_3$
    $$= \frac{(1.4 \text{ m}L)\,(0.02N)\,(50{,}000)}{100 \text{ m}L}$$

    $= 14$ mg/L as $CaCO_3$

    Total Alkalinity
    mg/L as $CaCO_3$
    $$= \frac{(2.8 \text{ m}L)\,(0.02N)\,(50{,}000)}{100 \text{ m}L}$$

    $= 28$ mg/L as $CaCO_3$

Use the alkalinity table to determine the remaining alkalinity constituents: (P = 1/2 T)

    Bicarbonate Alkalinity $= 0$ mg/L

    Carbonate Alkalinity $= 2P$

    $= 2\,(14$ mg/$L)$

    $= 28$ mg/L as $CaCO_3$

    Hydroxide Alkalinity $= 0$ mg/L

PRACTICE PROBLEMS 14.6

1. First calculate the A – D factors:

   $A = (CO_2, mg/L)(56/44)$

   $\quad = (7\ mg/L)(56/44)$

   $\quad = 9\ mg/L$

   $B = (Alkalinity, mg/L)(56/100)$

   $\quad = (140\ mg/L)(56/100)$

   $\quad = 78\ mg/L$

   $C = 0\ mg/L$

   $D = (Mg^{+2}, mg/L)(56/24.3)$

   $\quad = (25\ mg/L)(56/24.3)$

   $\quad = 58\ mg/L$

   Then calculate the estimated quicklime dosage:

   $$\text{Quicklime Dosage, } mg/L = \frac{(9\ mg/L + 78\ mg/L + 0 + 58\ mg/L)\ (1.15)}{0.90}$$

   $$= \frac{(145\ mg/L)\ (1.15)}{0.90}$$

   $$= 185\ mg/L\ CaO$$

2. Calculate the A – D factors:

   $A = (CO_2, mg/L)(74/44)$

   $\quad = (4\ mg/L)(74/44)$

   $\quad = 7\ mg/L$

   $B = (Alkalinity, mg/L)(74/100)$

   $\quad = (162\ mg/L)(74/100)$

   $\quad = 120\ mg/L$

   $C = 0\ mg/L$

   $D = (Mg^{+2}, mg/L)(74/24.3)$

   $\quad = (15\ mg/L)(74/24.3)$

   $\quad = 46\ mg/L$

   Then calculate the estimated hydrated lime dosage:

   $$\text{Hydrated Lime Dosage, } mg/L = \frac{(7\ mg/L + 120\ mg/L + 0 + 46\ mg/L)\ (1.15)}{0.90}$$

   $$= \frac{(173\ mg/L)\ (1.15)}{0.90}$$

   $$= 221\ mg/L\ Ca(OH)_2$$

3. First determine the A – D factors:

$A = (CO_2, mg/L)(74/44)$

$= (5 \ mg/L)(74/44)$

$= 8 \ mg/L$

$B = (Alkalinity, mg/L)(74/100)$

$= (108 \ mg/L)(74/100)$

$= 80 \ mg/L$

$C = 0 \ mg/L$

$D = (Mg^{+2}, mg/L)(74/24.3)$

$= (11 \ mg/L)(74/24.3)$

$= 33 \ mg/L$

Then calculate the estimated hydrated lime dosage:

$$\text{Hydrated Lime Dosage, } mg/L = \frac{(8 \ mg/L + 80 \ mg/L + 0 + 33 \ mg/L) \ (1.15)}{0.90}$$

$$= \frac{(121 \ mg/L) \ (1.15)}{0.90}$$

$$= 155 \ mg/L \ Ca(OH)_2$$

4. First calculate the A – D factors:

$A = (CO_2, mg/L)(56/44)$

$= (8 \ mg/L)(56/44)$

$= 10 \ mg/L$

$B = (Alkalinity, mg/L)(56/100)$

$= (170 \ mg/L)(56/100)$

$= 95 \ mg/L$

$C = 0 \ mg/L$

$D = (Mg^{+2}, mg/L)(56/24.3)$

$= (17 \ mg/L)(56/24.3)$

$= 39 \ mg/L$

Then calculate the estimated quicklime dosage:

$$\text{Quicklime Dosage, } mg/L = \frac{(10 \ mg/L + 95 \ mg/L + 0 + 39 \ mg/L) \ (1.15)}{0.90}$$

$$= \frac{(144 \ mg/L) \ (1.15)}{0.90}$$

$$= 184 \ mg/L \ CaO$$

PRACTICE PROBLEMS 14.7

1. First calculate the noncarbonate hardness:

$$\begin{array}{ccc} \text{Total Hardness,} & = & \text{Carbonate Hardness,} \\ \text{mg/}L \text{ as } CaCO_3 & & \text{mg/}L \text{ as } CaCO_3 \end{array} + \begin{array}{c} \text{Noncarbonate} \\ \text{Hardness} \\ \text{mg/}L \text{ as } CaCO_3 \end{array}$$

$$240 \text{ mg/}L = 173 \text{ mg/}L + x \text{ mg/}L$$

$$240 \text{ mg/}L - 173 \text{ mg/}L = x$$

$$67 \text{ mg/}L = x$$

Then calculate the soda ash required:

$$\begin{array}{c} \text{Soda Ash,} \\ \text{mg/}L \end{array} = \begin{array}{c} \text{(Noncarbonate)} \\ \text{Hardness} \\ \text{mg/}L \text{ as } CaCO_3 \end{array} \frac{(106)}{100}$$

$$= \frac{(67 \text{ mg/}L)(106)}{100}$$

$$= 71 \text{ mg/}L \text{ soda ash}$$

2. The noncarbonate hardness is:

$$225 \text{ mg/}L = 109 \text{ mg/}L + x \text{ mg/}L$$

$$225 \text{ mg/}L - 109 \text{ mg/}L = x$$

$$116 \text{ mg/}L = x$$

The soda ash required is:

$$\begin{array}{c} \text{Soda Ash,} \\ \text{mg/}L \end{array} = \begin{array}{c} \text{(Noncarbonate)} \\ \text{Hardness} \\ \text{mg/}L \text{ as } CaCO_3 \end{array} \frac{(106)}{100}$$

$$= \frac{(116 \text{ mg/}L)(106)}{100}$$

$$= 123 \text{ mg/}L \text{ soda ash}$$

3. The noncarbonate hardness is:

$$258 \text{ mg/}L = 160 \text{ mg/}L + x \text{ mg/}L$$

$$258 \text{ mg/}L - 160 \text{ mg/}L = x$$

$$98 \text{ mg/}L = x$$

And the soda ash required for softening is:

$$\begin{array}{c} \text{Soda Ash,} \\ \text{mg/}L \end{array} = \frac{(98 \text{ mg/}L)(106)}{100}$$

$$= 104 \text{ mg/}L \text{ soda ash}$$

4. First determine the noncarbonate hardness:

$$218 \text{ mg/L} = 110 \text{ mg/L} + x \text{ mg/L}$$

$$218 \text{ mg/L} - 110 \text{ mg/L} = x$$

$$108 \text{ mg/L} = x$$

And the soda ash required for softening is:

$$\text{Soda Ash, mg/L} = (108 \text{ mg/L}) \frac{(106)}{100}$$

$$= 114 \text{ mg/L soda ash}$$

PRACTICE PROBLEMS 14.8

1. First calculate the excess lime concentration:

$$\text{Excess Lime, mg/L} = (A + B + C + D)(0.15)$$

$$= (10 \text{ mg/L} + 132 \text{ mg/L} + 0 + 65 \text{ mg/L})(0.15)$$

$$= (207 \text{ mg/L})(0.15)$$

$$= 31 \text{ mg/L}$$

Then determine the required carbon dioxide dosage:

$$\text{Total CO}_2 \text{ Dosage, mg/L} = (31 \text{ mg/L})\frac{(44)}{74} + (5 \text{ mg/L})\frac{(44)}{24.3}$$

$$= 18 \text{ mg/L} + 9 \text{ mg/L}$$

$$= 27 \text{ mg/L CO}_2$$

2. The excess lime concentration is:

$$\text{Excess Lime, mg/L} = (A + B + C + D)(0.15)$$

$$= (9 \text{ mg/L} + 92 \text{ mg/L} + 6 \text{ mg/L} + 110 \text{ mg/L})(0.15)$$

$$= (217 \text{ mg/L})(0.15)$$

$$= 33 \text{ mg/L}$$

And the required carbon dioxide dosage for recarbonation is:

$$\text{Total CO}_2 \text{ Dosage, mg/L} = (33 \text{ mg/L})\frac{(44)}{74} + (4 \text{ mg/L})\frac{(44)}{24.3}$$

$$= 20 \text{ mg/L} + 7 \text{ mg/L}$$

$$= 27 \text{ mg/L CO}_2$$

PRACTICE PROBLEMS —Cont'd

3. The excess lime concentration is:

$$\text{Excess Lime,} \atop mg/L = (A + B + C + D)\ (0.15)$$

$$= (8\ mg/L + 111\ mg/L + 2\ mg/L + 54\ mg/L)\ (0.15)$$

$$= (175\ mg/L\ )\ (0.15)$$

$$= 26\ mg/L$$

And the required carbon dioxide dosage for recarbonation is:

$$\text{Total }CO_2 \atop \text{Dosage, } mg/L = (26\ mg/L)\ \frac{(44)}{74} + (6\ mg/L)\ \frac{(44)}{24.3}$$

$$= 15\ mg/L + 11\ mg/L$$

$$= 26\ mg/L\ CO_2$$

4. The excess lime is:

$$\text{Excess Lime,} \atop mg/L = (A + B + C + D)\ (0.15)$$

$$= (7\ mg/L + 115\ mg/L + 7\ mg/L + 47\ mg/L)\ (0.15)$$

$$= (169\ mg/L\ )\ (0.15)$$

$$= 26\ mg/L$$

And the required carbon dioxide dosage for recarbonation is:

$$\text{Total }CO_2 \atop \text{Dosage, } mg/L = (26\ mg/L)\ \frac{(44)}{74} + (3\ mg/L)\ \frac{(44)}{24.3}$$

$$= 15\ mg/L + 5\ mg/L$$

$$= 20\ mg/L\ CO_2$$

PRACTICE PROBLEMS 14.9

1. (mg/L Chemical) (MGD flow) (8.34 lbs/gal) = lbs/day Chemical

   (190 mg/L) (2.46 MGD) (8.34 lbs/gal) = 3898 lbs/day

2. First calculate the lbs/day feed rate:

   (mg/L Chemical) (MGD flow) (8.34 lbs/gal) = lbs/day Chemical

   (185 mg/L) (3.14 MGD) (8.34 lbs/gal) = 4845 lbs/day

   Then calculate the lbs/min feed rate:

$$\frac{4845\ lbs/day}{1440\ min/day} = 3.4\ lbs/min$$

3. First calculate the lbs/day feed rate:

   (mg/*L* Chemical) (MGD flow) (8.34 lbs/gal) = lbs/day Chemical

   (65 mg/*L*) (4.55 MGD) (8.34 lbs/gal) = 2467 lbs/day

   Then convert lbs/day to lbs/hr feed rate:

   $$\frac{2467 \text{ lbs/day}}{24 \text{ hrs/day}} = 103 \text{ lbs/hr}$$

4. The lbs/day feeder setting is:

   (mg/*L* Chemical) (MGD flow) (8.34 lbs/gal) = lbs/day Chemical

   (135 mg/*L*) (1.97 MGD) (8.34 lbs/gal) = 2218 lbs/day

   The lbs/min feed rate is:

   $$\frac{2218 \text{ lbs/day}}{1440 \text{ min/day}} = 1.5 \text{ lbs/min}$$

5. First calculate the lbs/day feed rate:

   (mg/*L* Chemical) (MGD flow) (8.34 lbs/gal) = lbs/day Chemical

   (40 mg/*L*) (3.05 MGD) (8.34 lbs/gal) = 1017 lbs/day

   Next, calculate the lbs/hr feed rate:

   $$\frac{1017 \text{ lbs/day}}{24 \text{ hrs/day}} = 43 \text{ lbs/hr}$$

   Then calculate the lbs/min feed rate:

   $$\frac{43 \text{ lbs/hr}}{60 \text{ min/hr}} = 0.7 \text{ lbs/min}$$

PRACTICE PROBLEMS 14.10

1. $\dfrac{216 \text{ mg/}L}{17.12 \text{ mg/}L/\text{gpg}} = 12.6 \text{ gpg}$

2. $(12.8 \text{ gpg}) \left(17.12 \dfrac{\text{mg/}L}{\text{gpg}}\right) = 219 \text{ mg/}L$

3. $\dfrac{245 \text{ mg/}L}{17.12 \text{ mg/}L/\text{gpg}} = 14.3 \text{ gpg}$

4. $(16 \text{ gpg}) \left(17.12 \dfrac{\text{mg/}L}{\text{gpg}}\right) = 274 \text{ mg/}L$

PRACTICE PROBLEMS 14.11

1. Exchange Capacity, grains

$$= \frac{(\text{Removal Capacity,})}{\text{grains/cu ft}}(\text{Media Volume,})\text{cu ft}$$

$$= (24{,}000 \text{ grains/cu ft}) (110 \text{ cu ft})$$

$$= 2{,}640{,}000 \text{ grains}$$

2. First calculate the cu ft volume of resin:

$$\text{Vol., cu ft} = (0.785)(D^2)(\text{Depth, ft})$$

$$= (0.785)(5 \text{ ft})(5 \text{ ft})(3.5 \text{ ft})$$

$$= 69 \text{ cu ft}$$

Now calculate the exchange capacity of the softener:

Exchange Capacity, grains

$$= \frac{(\text{Removal Capacity,})}{\text{grains/cu ft}}(\text{Media Volume,})\text{cu ft}$$

$$= (20{,}000 \text{ grains/cu ft})(69 \text{ cu ft})$$

$$= 1{,}380{,}000 \text{ grains}$$

3. First convert kilograins to grains:

$$22 \text{ kilograins} = 22{,}000 \text{ grains}$$

Then calculate the exchange capacity of the softener:

Exchange Capacity, grains

$$= \frac{(\text{Removal Capacity,})}{\text{grains/cu ft}}(\text{Media Volume,})\text{cu ft}$$

$$= (22{,}000 \text{ grains/cu ft})(280 \text{ cu ft})$$

$$= 6{,}160{,}000 \text{ grains}$$

4. The cu ft volume of resin is:

$$\text{Vol., cu ft} = (0.785)(D^2)(\text{Depth, ft})$$

$$= (0.785)(6 \text{ ft})(6 \text{ ft})(4 \text{ ft})$$

$$= 113 \text{ cu ft}$$

The exchange capacity can now be calculated:

Exchange Capacity, grains

$$= \frac{(\text{Removal Capacity,})}{\text{grains/cu ft}}(\text{Media Volume,})\text{cu ft}$$

$$= (20{,}000 \text{ grains/cu ft})(113 \text{ cu ft})$$

$$= 2{,}260{,}000 \text{ grains}$$

PRACTICE PROBLEMS 14.12

1. $\dfrac{\text{Water Treatment}}{\text{Capacity, gal}} = \dfrac{\text{Exchange Capacity, grains}}{\text{Hardness, grains/gallon}}$

   $= \dfrac{2,192,000 \text{ grains}}{17.8 \text{ gpg}}$

   $= 123,146 \text{ gallons}$

2. $\dfrac{\text{Water Treatment}}{\text{Capacity, gal}} = \dfrac{\text{Exchange Capacity, grains}}{\text{Hardness, grains/gallon}}$

   $= \dfrac{4,500,000 \text{ grains}}{16.2 \text{ gpg}}$

   $= 277,778 \text{ gallons}$

3. First, convert the mg/L hardness to gpg hardness:

   $$\frac{265 \text{ mg/}L}{17.12 \text{ mg/}L \text{ /gpg}} = 15.5 \text{ gpg}$$

   Now the gallons water treated can be calculated:

   $\dfrac{\text{Water Treatment}}{\text{Capacity, gal}} = \dfrac{\text{Exchange Capacity, grains}}{\text{Hardness, grains/gallon}}$

   $= \dfrac{3,877,000 \text{ grains}}{15.5 \text{ gpg}}$

   $= 250,129 \text{ gallons}$

4. The exchange capacity of the softener must be calculated:

   $\dfrac{\text{Exchange Capacity,}}{\text{grains}} = \dfrac{(\text{Removal Capac.,}) (\text{Media Vol.,})}{\text{grains/cu ft} \qquad \text{cu ft}}$

   $= (20,000 \text{ grains/cu ft}) (180 \text{ cu ft})$

   $= 3,600,000 \text{ grains}$

   The water treated before regeneration can now be determined:

   $\dfrac{\text{Water Treatment}}{\text{Capacity, gal}} = \dfrac{3,600,000 \text{ grains}}{14.9 \text{ gpg}}$

   $= 241,611 \text{ gallons}$

PRACTICE PROBLEMS—Cont'd

5. First calculate the cu ft volume of resin:

$$\text{Vol., cu ft} = (0.785)(D^2)(\text{Depth, ft})$$

$$= (0.785)(4 \text{ ft})(4 \text{ ft})(2.5 \text{ ft})$$

$$= 31.4 \text{ cu ft}$$

Then calculate the exchange capacity of the softener:

$$\text{Exchange Capacity, grains} = (\text{Removal Capacity, grains/cu ft})(\text{Media Volume, cu ft})$$

$$= (21,000 \text{ grains/cu ft})(31.4 \text{ cu ft})$$

$$= 659,400 \text{ grains}$$

The gallons of water that can be treated before regeneration are:

$$\text{Water Treatment Capacity, gal} = \frac{659,400 \text{ grains}}{13.7 \text{ gpg}}$$

$$= 48,131 \text{ gallons}$$

PRACTICE PROBLEMS 14.13

1. $$\text{Operating Time, hrs} = \frac{\text{Water Treated, gal}}{\text{Flow Rate, gph}}$$

$$= \frac{590,000 \text{ gal}}{24,500 \text{ gph}}$$

$$= 24.1 \text{ hrs of operation}$$

2. $$\text{Operating Time, hrs} = \frac{\text{Water Treated, gal}}{\text{Flow Rate, gph}}$$

$$= \frac{782,000 \text{ gal}}{25,000 \text{ gph}}$$

$$= 31.3 \text{ hrs of operation}$$

3. Since the operating time is desired in hours, the flow rate should be expressed in terms of hours as well:

$$(225 \text{ gpm})(60 \text{ min/hr}) = 13,500 \text{ gph}$$

The operating time required is:

$$\text{Operating Time, hrs} = \frac{\text{Water Treated, gal}}{\text{Flow Rate, gph}}$$

$$= \frac{356,000 \text{ gal}}{13,500 \text{ gph}}$$

$$= 26.4 \text{ hrs of operating time}$$

4. First calculate the total gallons treated before regeneration:

$$\text{Water Treatment, gal} = \frac{\text{Exchange Capacity, grains}}{\text{Hardness, gpg}}$$

$$= \frac{3{,}158{,}000 \text{ grains}}{13 \text{ gpg}}$$

$$= 242{,}923 \text{ gallons}$$

And since the operating time is required in hours, the flow rate should be expressed as gph:

$$(210 \text{ gpm})(60 \text{ min/hr}) = 12{,}600 \text{ gph}$$

The operating time may now be calculated:

$$\text{Operating Time, hrs} = \frac{\text{Water Treated, gal}}{\text{Flow Rate, gph}}$$

$$= \frac{242{,}923 \text{ gal}}{12{,}600 \text{ gph}}$$

$$= 19.3 \text{ hrs operating time}$$

5. First calculate the total gallons treated before regeneration:

$$\text{Water Treatment, gal} = \frac{\text{Exchange Capacity, grains}}{\text{Hardness, gpg}}$$

$$= \frac{3{,}780{,}000 \text{ grains}}{11.4 \text{ gpg}}$$

$$= 331{,}579 \text{ gallons}$$

The flow rate should be expressed as gph, since the operating time is desired in hours:

$$\frac{289{,}000 \text{ gpd}}{24 \text{ hrs/day}} = 12{,}042 \text{ gph}$$

The operating time may now be calculated:

$$\text{Operating Time, hrs} = \frac{\text{Water Treated, gal}}{\text{Flow Rate, gph}}$$

$$= \frac{331{,}579 \text{ gal}}{12{,}042 \text{ gph}}$$

$$= 27.5 \text{ hrs operating}$$

PRACTICE PROBLEMS 14.14

1. Salt Required, $= \dfrac{(\text{Salt Req'd,})\ (\text{Hardness Removed,})}{\text{lbs/1000 grains} \quad \text{grains}}$
   lbs

   $= \dfrac{(0.4\ \text{lbs salt})\ (2300\ \text{kilograins})}{\text{kilograins rem.}}$

   $= 920$ lbs salt required

2. Salt Required, $= \dfrac{(\text{Salt Req'd,})\ (\text{Hardness Removed,})}{\text{lbs/1000 grains} \quad \text{grains}}$
   lbs

   $= \dfrac{(0.3\ \text{lbs salt})\ (1330\ \text{kilograins})}{\text{kilograins rem.}}$

   $= 399$ lbs salt required

3. Brine, $= \dfrac{\text{Salt Required, lbs}}{\text{Salt Solution, lbs salt/gal brine}}$
   gal

   $= \dfrac{395\ \text{lbs salt}}{1.19\ \text{lbs salt/gal brine}}$

   $= 332$ gal of 13% brine

4. Brine, $= \dfrac{\text{Salt Required, lbs}}{\text{Salt Solution, lbs salt/gal brine}}$
   gal

   $= \dfrac{410\ \text{lbs salt}}{1.29\ \text{lbs salt/gal brine}}$

   $= 318$ gal of 14% brine

5. First calculate the lbs salt required:

   Salt Required, $= \dfrac{(\text{Salt Req'd,})\ (\text{Hardness Removed,})}{\text{lbs/1000 grains} \quad \text{grains}}$
   lbs

   $= \dfrac{(0.4\ \text{lbs salt})\ (1310\ \text{kilograins})}{\text{kilograins rem.}}$

   $= 524$ lbs salt required

   Then calculate the gallons brine required:

   Brine, $= \dfrac{\text{Salt Required, lbs}}{\text{Salt Solution, lbs salt/gal brine}}$
   gal

   $= \dfrac{524\ \text{lbs salt}}{1.09\ \text{lbs salt/gal brine}}$

   $= 481$ gal of 12% brine

CHAPTER 14 ACHIEVEMENT TEST

1. $\begin{aligned} \text{Equivalent} \\ \text{Weight of Na} \end{aligned} = \dfrac{22.997}{1}$

$= 22.997$

2. Calcium Hardness,

$$\frac{\text{mg/}L \text{ as CaCO}_3}{\text{Equivalent Wt. of CaCO}_3} = \frac{\text{Calcium, mg/}L}{\text{Equivalent Weight of Calcium}}$$

$$\frac{x \text{ mg/}L}{50.045} = \frac{42 \text{ mg/}L}{20.04}$$

$$x \text{ mg/}L = \frac{(42 \text{mg/}L)\,(50.045)}{20.04}$$

$$x = 105 \text{ mg/}L \text{ Ca} \\ \text{as CaCO}_3$$

3. $\begin{aligned} \text{Phenolphthalein} \\ \text{Alkalinity} \\ \text{mg/}L \text{ as CaCO}_3 \end{aligned} = \dfrac{(A)(N)(50{,}000)}{\text{m}L \text{ of Sample}}$

$$= \frac{(1.6 \text{ m}L)\,(0.02N)\,(50{,}000)}{100 \text{ m}L}$$

$$= 16 \text{ mg/}L \text{ as CaCO}_3$$

4. P alkalinity (10 mg/L) is less than half of the alkalinity (52 mg/L ÷ 2 = 26 mg/L)

Bicarbonate Alkalinity $= T - 2P$

$$= 52 \text{ mg/}L - 2\,(10 \text{ mg/}L)$$

$$= 52 \text{ mg/}L - 20 \text{ mg/}L$$

$$= 32 \text{ mg/}L \text{ as CaCO}_3$$

Carbonate Alkalinity $= 2P$

$$= 2\,(10 \text{ mg/}L)$$

$$= 20 \text{ mg/}L \text{ as CaCO}_3$$

Hydroxide Alkalinity $= 0$ mg/L

ACHIEVEMENT TEST—Cont'd

5. First calculate the molecular weight of $MgCl_2$:

$$Mg:\ 24.32 \times 1 = 24.32$$
$$Cl:\ 35.957 \times 2 = \underline{71.914}$$
$$96.234\ \text{Molecular Weight}$$

Then calculate the equivalent weight of $MgCl_2$: (The valence of Mg is 2.)

$$\text{Equivalent Weight of } MgCl_2 = \frac{\text{Molecular Weight}}{\text{Net Positive Valence}}$$

$$= \frac{96.234}{2}$$

$$= 48.117$$

6.

$$\frac{\text{Magnesium Hardness, } mg/L \text{ as } CaCO_3}{\text{Equivalent Wt. of } CaCO_3} = \frac{\text{Magnesium, } mg/L}{\text{Equivalent Wt of Magnesium}}$$

$$\frac{x\ mg/L}{50.045} = \frac{28\ mg/L}{12.16}$$

$$x = \frac{(28)\ (50.045)}{12.16}$$

$$x = 115\ mg/L \text{ as } CaCO_3$$

7.

$$\text{Number of Milliequivalents} = \frac{\text{Chemical, } mg}{\text{Equivalent Wt.}}$$

$$= \frac{22\ mg}{12.16}$$

$$= 1.8\ \text{milliequivalents}$$

8. First determine the A – D factors:

$$A = (CO_2,\ mg/L)(74/44)$$
$$= (7\ mg/L)(74/44)$$
$$= 12\ mg/L$$

$$B = (\text{Alkalinity, } mg/L)(74/100)$$
$$= (116\ mg/L)(74/100)$$
$$= 86\ mg/L$$

$$C = 0\ mg/L$$

$$D = (Mg^{+2},\ mg/L)(74/24.3)$$
$$= (14\ mg/L)(74/24.3)$$
$$= 43\ mg/L$$

Then calculate the estimated hydrated lime dosage:

$$\text{Hydrated Lime Dosage, } mg/L = \frac{(12\ mg/L + 86\ mg/L + 0 + 43\ mg/L)\ (1.15)}{0.90}$$

$$= \frac{(141\ mg/L)\ (1.15)}{0.90}$$

$$= 180\ mg/L\ Ca(OH)_2$$

9. First express the calcium and magnesium hardness in mg/*L* as $CaCO_3$:

   Calcium hardness: Magnesium hardness:

   $$\frac{x \text{ mg}/L}{50.045} = \frac{30 \text{ mg}/L}{20.04} \qquad\qquad \frac{x \text{ mg}/L}{50.045} = \frac{12 \text{ mg}/L}{12.16}$$

   $$x = 75 \text{ mg}/L \text{ Ca} \qquad\qquad x = 49 \text{ mg}/L \text{ Mg}$$
   $$\text{as } CaCO_3 \qquad\qquad\qquad \text{as } CaCO_3$$

   Now the total hardness can be calculated:

   $$\begin{array}{l}\text{Total Hardness,} \\ \text{mg}/L \text{ as } CaCO_3\end{array} = 75 \text{ mg}/L + 49 \text{ mg}/L$$

   $$= 124 \text{ mg}/L \text{ total hardness}$$
   $$\text{as } CaCO_3$$

10. The alkalinity (115 mg/*L* as $CaCO_3$) is greater than the total hardness (103 mg/*L* as $CaCO_3$). Therefore, all the hardness is carbonate hardness.

    $$\begin{array}{l}\text{Total Hardness,} \\ \text{mg}/L \text{ as } CaCO_3\end{array} = \begin{array}{l}\text{Carbonate Hardness,} \\ \text{mg}/L \text{ as } CaCO_3\end{array}$$

    $$\begin{array}{l}103 \text{ mg}/L \\ \text{as } CaCO_3\end{array} = \begin{array}{l}\text{Carbonate} \\ \text{Hardness}\end{array}$$

11. The A – D factors are:

    $$A = (CO_2 , \text{mg}/L)(56/44)$$
    $$= (4 \text{ mg}/L)(56/44)$$
    $$= 5 \text{ mg}/L$$
    $$B = (\text{Alkalinity, mg}/L)(56/100)$$
    $$= (155 \text{ mg}/L)(56/100)$$
    $$= 87 \text{ mg}/L$$

    $$C = 0 \text{ mg}/L$$

    $$D = (Mg^{+2}, \text{mg}/L)(56/24.3)$$
    $$= (12 \text{ mg}/L)(56/24.3)$$
    $$= 28 \text{ mg}/L$$

    The estimated quicklime dosage can now be calculated:

    $$\begin{array}{l}\text{Quicklime Dosage,} \\ \text{mg}/L\end{array} = \frac{(5 \text{ mg}/L + 87 \text{ mg}/L + 0 + 28 \text{ mg}/L) (1.15)}{0.90}$$

    $$= \frac{(120 \text{ mg}/L) (1.15)}{0.90}$$

    $$= 153 \text{ mg}/L \text{ CaO}$$

12. $$\begin{array}{l}\text{Phenolphthalein} \\ \text{Alkalinity} \\ \text{mg}/L \text{ as } CaCO_3\end{array} = \frac{(A)(N)(50,000)}{\text{m}L \text{ of Sample}}$$

    $$= \frac{(0.4 \text{ m}L) (0.02N) (50,000)}{100 \text{ m}L}$$

    $$= \begin{array}{l}4 \text{ mg}/L \text{ as } CaCO_3 \\ \text{Phenolphthalein Alkalinity}\end{array}$$

ACHIEVEMENT TEST—Cont'd

$$\text{Total Alkalinity,} \atop \text{mg/}L \text{ as } CaCO_3 = \frac{(6.7 \text{ m}L)(0.02N)(50{,}000)}{100 \text{ m}L}$$

$$= 67 \text{ mg/}L \text{ as } CaCO_3$$

13. $$\text{Total Hardness,} \atop \text{mg/}L \text{ as } CaCO_3 = \text{Carbonate Hardness,} \atop \text{mg/}L \text{ as } CaCO_3 + \text{Noncarbonate Hardness,} \atop \text{mg/}L \text{ as } CaCO_3$$

$$118 \text{ mg/}L = 85 \text{ mg/}L + x \text{ mg/}L$$

$$118 \text{ mg/}L - 85 \text{ mg/}L = x \text{ mg/}L$$

$$33 \text{ mg/}L = x$$
noncarbonate
hardness

(Carbonate hardness is 85 mg/L, as indicated above.)

14. First calculate the noncarbonate hardness:

$$214 \text{ mg/}L = 95 \text{ mg/}L + x \text{ mg/}L$$

$$214 \text{ mg/}L - 95 \text{ mg/}L = x$$

$$119 \text{ mg/}L = x$$

The soda ash required is:

$$\text{Soda Ash,} \atop \text{mg/}L = \text{(Noncarbonate)} \atop \text{Hardness} \atop \text{mg/}L \text{ as } CaCO_3 \frac{(106)}{100}$$

$$= \frac{(119 \text{ mg/}L)(106)}{100}$$

$$= 126 \text{ mg/}L \text{ soda ash}$$

15. P alkalinity (15 mg/L) is greater than 1/2 T alkalinity (24 mg/L ÷ 2 = 12 mg/L):

Bicarbonate Alkalinity $= 0$ mg/L

Carbonate Alkalinity $= 2T - 2P$

$$= 2 (24 \text{ mg/}L) - 2 (15 \text{ mg/}L)$$

$$= 48 \text{ mg/}L - 30 \text{ mg/}L$$

$$= 18 \text{ mg/}L \text{ as } CaCO_3$$

Hydroxide Alkalinity $= 2P - T$

$$= 2 (15 \text{ mg/}L) - 24 \text{ mg/}L$$

$$= 6 \text{ mg/}L$$

16. Exchange Capacity, grains $= \dfrac{\text{(Removal Capacity,)(Media Volume,)}}{\text{grains/cu ft} \qquad \text{cu ft}}$

$= (21{,}000 \text{ grains/cu ft}) (90 \text{ cu ft})$

$= 1{,}890{,}000 \text{ grains}$

17. The noncarbonate hardness is:

$$255 \text{ mg/}L = 158 \text{ mg/}L + x \text{ mg/}L$$

$$255 \text{ mg/}L - 158 \text{ mg/}L = x$$

$$97 \text{ mg/}L = x$$

And the soda ash required for softening is:

$$\dfrac{\text{Soda Ash,}}{\text{mg/}L} = (97 \text{ mg/}L) \dfrac{(106)}{100}$$

$$= 103 \text{ mg/}L \text{ soda ash}$$

18. $(14 \text{ gpg}) (17.12 \text{ mg/}L\text{/gpg}) = 240 \text{ mg/}L$

19. First calculate the excess lime concentration:

$$\dfrac{\text{Excess Lime,}}{\text{mg/}L} = (A + B + C + D) (0.15)$$

$$= (8 \text{ mg/}L + 140 \text{ mg/}L + 3 \text{ mg/}L + 63 \text{ mg/}L) (0.15)$$

$$= (214 \text{ mg/}L) (0.15)$$

$$= 32 \text{ mg/}L$$

The required carbon dioxide dosage can now be determined:

$$\dfrac{\text{Total CO}_2}{\text{Dosage, mg/}L} = (32 \text{ mg/}L) \dfrac{(44)}{74} + (7 \text{ mg/}L) \dfrac{(44)}{24.3}$$

$$= 19 \text{ mg/}L + 13 \text{ mg/}L$$

$$= 32 \text{ mg/}L \text{ CO}_2$$

20. Water Treatment Capacity, gal $= \dfrac{\text{Exchange Capacity, grains}}{\text{Hardness, grains/gallon}}$

$$= \dfrac{2{,}207{,}000 \text{ grains}}{17.6 \text{ gpg}}$$

$$= 125{,}398 \text{ gallons}$$

ACHIEVEMENT TEST—Cont'd

21. The cu ft volume of resin is:

$$\text{Vol., cu ft} = (0.785)(D^2)(\text{Depth, ft})$$

$$= (0.785)(6 \text{ ft})(6 \text{ ft})(4 \text{ ft})$$

$$= 113 \text{ cu ft}$$

The exchange capacity can now be calculated:

$$\text{Exchange Capacity, grains} = (\text{Removal Capacity, grains/cu ft})(\text{Media Volume, cu ft})$$

$$= (19,000 \text{ grains/cu ft})(113 \text{ cu ft})$$

$$= 2,147,000 \text{ grains}$$

22. The excess lime concentration is:

$$\text{Excess Lime, mg/L} = (A + B + C + D)(0.15)$$

$$= (8 \text{ mg/L} + 90 \text{ mg/L} + 7 \text{ mg/L} + 105 \text{ mg/L})(0.15)$$

$$= (210 \text{ mg/L})(0.15)$$

$$= 32 \text{ mg/L}$$

The required carbon dioxide dosage can now be determined:

$$\text{Total CO}_2 \text{ Dosage, mg/L} = (32 \text{ mg/L})\frac{(44)}{74} + (5 \text{ mg/L})\frac{(44)}{24.3}$$

$$= 19 \text{ mg/L} + 9 \text{ mg/L}$$

$$= 28 \text{ mg/L CO}_2$$

23. The ion exchange capacity of the softener must be calculated:

$$\text{Exchange Capacity, grains} = (\text{Removal Capacity, grains/cu ft})(\text{Media Volume, cu ft})$$

$$= (21,000 \text{ grains/cu ft})(190 \text{ cu ft})$$

$$= 3,990,000 \text{ grains}$$

The water treated before regeneration can now be determined:

$$\text{Water Treatment Capacity, gal} = \frac{\text{Exchange Capacity, grains}}{\text{Hardness, grains/gallon}}$$

$$= \frac{3,990,000 \text{ grains}}{15.1 \text{ gpg}}$$

$$= 264,238 \text{ gallons}$$

24. First calculate the lbs/day feed rate:

(mg/*L* Chemical) (MGD flow) (8.34 lbs/gal) = lbs/day Chemical

(170 mg/*L*) (2.065 MGD) (8.34 lbs/gal) = 2928 lbs/day

Then calculate the lbs/min feed rate:

$$\frac{2928 \text{ lbs/day}}{1440 \text{ min/day}} = 2 \text{ lbs/min}$$

25. $$\frac{229 \text{ mg/}L}{17.12 \text{ mg/}L\text{/gpg}} = 13.4 \text{ gpg}$$

26. $$\frac{\text{Operating}}{\text{Time, hrs}} = \frac{\text{Water Treated, gal}}{\text{Flow Rate, gph}}$$

$$= \frac{583,000 \text{ gal}}{24,720 \text{ gph}}$$

$$= 23.6 \text{ hrs of operation}$$

27. $$\frac{\text{Salt Required,}}{\text{lbs}} = \frac{(\text{Salt Req'd, })}{\text{lbs/1000 grains}} \frac{(\text{Hardness Removed,})}{\text{grains}}$$

$$= \frac{(\text{ 0.3 lbs salt })\;(2268 \text{ kilograins})}{\text{kilograins rem.}}$$

$$= 680 \text{ lbs salt required}$$

28. First calculate the lbs/day feed rate:

(mg/*L* Chemical) (MGD flow) (8.34 lbs/gal) = lbs/day Chemical

(60 mg/*L*) (3.15 MGD) (8.34 lbs/gal) = 1576 lbs/day

Then convert lbs/day to lbs/hr feed rate:

$$\frac{1576 \text{ lbs/day}}{24 \text{ hrs/day}} = 66 \text{ lbs/hr}$$

29. $$\frac{\text{Brine,}}{\text{gal}} = \frac{\text{Salt Required, lbs}}{\text{Salt Solution, lbs salt/gal brine}}$$

$$= \frac{409 \text{ lbs salt}}{0.990 \text{ lbs salt/gal brine}}$$

$$= 413 \text{ gal of brine}$$

ACHIEVEMENT TEST—Cont'd

30. Since the operating time is desired in hours, the flow rate should be expressed in hours as well:

$$(230 \text{ gpm}) (60 \text{ min/hr}) = 13,800 \text{ gph}$$

The required operating time is:

$$\frac{\text{Operating}}{\text{Time, hrs}} = \frac{\text{Water Treated, gal}}{\text{Flow Rate, gph}}$$

$$= \frac{352,000 \text{ gal}}{13,800 \text{ gph}}$$

$$= 25.5 \text{ hrs operating time}$$

Chapter 15

PRACTICE PROBLEMS 15.1

1. Since flow rate is desired in minutes, the time should be expressed in minutes as well:

$$\frac{72 \text{ seconds}}{60 \text{ sec/min}} = 1.2 \text{ min}$$

Now calculate flow rate:

$$\frac{\text{Flow,}}{\text{gpm}} = \frac{\text{Volume, gal}}{\text{Time, min}}$$

$$= \frac{1 \text{ gal}}{1.2 \text{ min}}$$

$$= 0.8 \text{ gpm}$$

2. First convert 56 seconds to minutes:

$$\frac{56 \text{ seconds}}{60 \text{ sec/min}} = 0.9 \text{ min}$$

Then calculate flow rate:

$$\frac{\text{Flow,}}{\text{gpm}} = \frac{\text{Volume, gal}}{\text{Time, min}}$$

$$= \frac{1 \text{ gal}}{0.9 \text{ min}}$$

$$= 1.1 \text{ gpm}$$

3. The time frame must be expressed in terms of minutes only:

$$\frac{38 \text{ seconds}}{60 \text{ sec/min}} = 0.6 \text{ min}$$

The total time is therefore 1.6 min. Now calculate the flow rate:

$$\text{Flow Rate, gpm} = \frac{\text{Volume, gal}}{\text{Time, min}}$$

$$= \frac{1 \text{ gal}}{1.6 \text{ min}}$$

$$= 0.6 \text{ gpm}$$

4. Use the same equation for flow rate, fill in the known factors, then solve for the unknown value: (First solve for <u>minutes</u>, then convert to <u>seconds</u>.)

$$\text{Flow, gpm} = \frac{\text{Volume, gal}}{\text{Time, min}}$$

$$0.6 \text{ gpm} = \frac{1 \text{ gal}}{x \text{ min}}$$

$$x = \frac{1}{0.6}$$

$$x = 1.7 \text{ min}$$

Now convert minutes to seconds:

$$(1.7 \text{ min}) (60 \text{ sec/min}) = 102 \text{ sec}$$

5. Use the flow rate equation, fill in the known factors, then solve for the unknown value: (First solve for <u>minutes</u>, then convert to <u>seconds</u>.)

$$\text{Flow, gpm} = \frac{\text{Volume, gal}}{\text{Time, min}}$$

$$0.4 \text{ gpm} = \frac{1 \text{ gal}}{x \text{ min}}$$

$$x = \frac{1}{0.4}$$

$$x = 2.5 \text{ min}$$

Now convert the minutes to seconds:

$$(2.5 \text{ min}) (60 \text{ sec/min}) = 150 \text{ sec}$$

PRACTICE PROBLEMS 15.2

1. First convert the pipe diameter to feet:

$$\frac{(0.75 \text{ inches})}{12 \text{ inches/ft}} = 0.06 \text{ ft}$$

Then calculate flushing time:

$$\begin{aligned}
\frac{\text{Flushing}}{\text{Time, min}} &= \frac{(0.785)(D^2)(\text{Length, ft})(7.48 \text{ gal/cu ft})(2)}{\text{Flow Rate, gpm}} \\
&= \frac{(0.785)(0.06 \text{ ft})(0.06 \text{ ft})(30 \text{ ft})(7.48 \text{ gal/cu ft})(2)}{0.5 \text{ gpm}} \\
&= 2.5 \text{ gpm}
\end{aligned}$$

2. The 3/4-inch pipe diameter is equivalent to a 0.06 ft diameter.

$$\begin{aligned}
\frac{\text{Flushing}}{\text{Time, min}} &= \frac{(0.785)(D^2)(\text{Length, ft})(7.48 \text{ gal/cu ft})(2)}{\text{Flow Rate, gpm}} \\
&= \frac{(0.785)(0.06 \text{ ft})(0.06 \text{ ft})(25 \text{ ft})(7.48 \text{ gal/cu ft})(2)}{0.6 \text{ gpm}} \\
&= 1.8 \text{ min}
\end{aligned}$$

3. 3/4-inch diameter = 0.06 ft diameter

$$\begin{aligned}
\frac{\text{Flushing}}{\text{Time, min}} &= \frac{(0.785)(D^2)(\text{Length, ft})(7.48 \text{ gal/cu ft})(2)}{\text{Flow Rate, gpm}} \\
&= \frac{(0.785)(0.06 \text{ ft})(0.06 \text{ ft})(35 \text{ ft})(7.48 \text{ gal/cu ft})(2)}{0.5 \text{ gpm}} \\
&= 3 \text{ min}
\end{aligned}$$

4. The 1/2-inch diameter line must be expressed in terms of feet:

$$\frac{(0.5 \text{ in.})}{12 \text{ in./ft}} = 0.04 \text{ ft}$$

The flushing time is therefore can now be calculated:

$$\text{Flushing Time, min} = \frac{(0.785)(D^2)(\text{Length, ft})(7.48 \text{ gal/cu ft})(2)}{\text{Flow Rate, gpm}}$$

$$= \frac{(0.785)(0.04 \text{ ft})(0.04 \text{ ft})(20 \text{ ft})(7.48 \text{ gal/cu ft})(2)}{0.5 \text{ gpm}}$$

$$= 0.8 \text{ min}$$

Convert to seconds:

$$(0.8 \text{ min})(60 \text{ sec/min}) = 48 \text{ seconds}$$

PRACTICE PROBLEMS 15.3

1. $\text{Normality} = \dfrac{\text{No. of Equivalents of Solute}}{\text{Liters of Solution}}$

$$= \frac{2.3 \text{ Equivalents}}{1.5 \text{ liters}}$$

$$= 1.5 \, N$$

2. First convert 700 mL to liters:

$$\frac{700 \text{ mL}}{1000 \text{ mL/L}} = 0.7 \, L$$

Then calculate the normality of the solution:

$$\text{Normality} = \frac{\text{No. of Equivalents of Solute}}{\text{Liters of Solution}}$$

$$= \frac{1.2 \text{ Equivalents}}{0.7 \text{ Liters}}$$

$$= 1.7 \text{ N}$$

3. $N_A V_A = N_B V_B$

$$(0.5)(x \text{ mL}) = (0.02)(400 \text{ mL})$$

$$x = \frac{(0.02)(400)}{0.5}$$

$$= 16 \text{ mL NaOH}$$

PRACTICE PROBLEMS—Cont'd

4. $$\frac{\text{Desired Wt., grams}}{\text{Desired Sol'n Vol., mL}} = \frac{\text{Actual Wt, grams}}{\text{Actual Sol'n Vol., mL}}$$

$$\frac{5 \text{ grams}}{1000 \text{ mL}} = \frac{4.72 \text{ grams}}{x \text{ mL}}$$

$$(5)(x) = (4.72)(1000)$$

$$x = \frac{(4.72)(1000)}{5}$$

$$x = 944 \text{ mL}$$

5. $$\frac{\text{Desired Wt., grams}}{\text{Desired Sol'n Vol., mL}} = \frac{\text{Actual Wt, grams}}{\text{Actual Sol'n Vol., mL}}$$

$$\frac{100 \text{ mg}}{100 \text{ mL}} = \frac{97.24 \text{ mg}}{x \text{ mL}}$$

$$(100)(x) = (97.24)(100)$$

$$x = \frac{(97.24)(100)}{100}$$

$$x = 97.24 \text{ mL}$$

6. $$\frac{\text{Desired Wt., grams}}{\text{Desired Sol'n Vol., mL}} = \frac{\text{Actual Wt, grams}}{\text{Actual Sol'n Vol., mL}}$$

$$\frac{100 \text{ grams}}{1000 \text{ mL}} = \frac{93.42 \text{ grams}}{x \text{ mL}}$$

$$(100)(x) = (93.42)(1000)$$

$$x = \frac{(93.42)(1000)}{100}$$

$$x = 934 \text{ mL}$$

PRACTICE PROBLEMS 15.4

1. $\text{Estim. Cl}_2 \text{ Resid., mg/}L = \dfrac{(\text{Cl}_2 \text{ Resid., mg/}L)(\text{Distilled } H_2O, mL)}{(\text{Sample Vol., drops})(0.05 \, mL/\text{drop})}$

    $= \dfrac{(0.4 \, \text{mg/}L)(10 \, mL)}{(3 \, \text{drops})(0.05 \, mL/\text{drop})}$

    $= 27 \, \text{mg/}L$

2. $\text{Estim. Cl}_2 \text{ Resid., mg/}L = \dfrac{(\text{Cl}_2 \text{ Resid., mg/}L)(\text{Distilled } H_2O, mL)}{(\text{Sample Vol., drops})(0.05 \, mL/\text{drop})}$

    $= \dfrac{(0.3 \, \text{mg/}L)(10 \, mL)}{(2 \, \text{drops})(0.05 \, mL/\text{drop})}$

    $= 30 \, \text{mg/}L$

3. $\text{Estim. Cl}_2 \text{ Resid., mg/}L = \dfrac{(\text{Cl}_2 \text{ Resid., mg/}L)(\text{Distilled } H_2O, mL)}{(\text{Sample Vol., drops})(0.05 \, mL/\text{drop})}$

    $= \dfrac{(0.2 \, \text{mg/}L)(10 \, mL)}{(3 \, \text{drops})(0.05 \, mL/\text{drop})}$

    $= 13 \, \text{mg/}L$

4. $\text{Estim. Cl}_2 \text{ Resid., mg/}L = \dfrac{(\text{Cl}_2 \text{ Resid., mg/}L)(\text{Distilled } H_2O, mL)}{(\text{Sample Vol., drops})(0.05 \, mL/\text{drop})}$

    $= \dfrac{(0.4 \, \text{mg/}L)(10 \, mL)}{(4 \, \text{drops})(0.05 \, mL/\text{drop})}$

    $= 20 \, \text{mg/}L$

PRACTICE PROBLEMS 15.5

1. Step 1: $68° + 40° = 108°$

    Step 2: $\dfrac{(5)(108°)}{9} = 60°$

    Step 3: $60° - 40° = 20° \, C$

2. Step 1: $14° + 40° = 54°$

    Step 2: $\dfrac{(9)(54°)}{5} = 97.2°$

    Step 3: $97.2° - 40° = 57.2° \, F$

PRACTICE PROBLEMS—Cont'd

3. Step 1: $56° + 40° = 96°$

   Step 2: $\dfrac{(5)}{9}(96°) = 53.3°$

   Step 3: $53.3° - 40° = 13.3°$ C

4. Step 1: $11° + 40° = 51°$

   Step 2: $\dfrac{(9)}{5}(51°) = 91.8°$

   Step 3: $91.8° - 40° = 51.8°$ F

CHAPTER 15 ACHIEVEMENT TEST

1. $\text{Flushing Time, min} = \dfrac{(0.785)(D^2)(\text{Length, ft})(7.48 \text{ gal/cu ft})(2)}{\text{Flow Rate, gpm}}$

    $= \dfrac{(0.785)(0.06 \text{ ft})(0.06 \text{ ft})(50 \text{ ft})(7.48 \text{ gal/cu ft})(2)}{0.75 \text{ gpm}}$

    $= 2.8 \text{ min}$

2. Since flow rate is desired in <u>minutes</u> the time should also be expressed as <u>minutes</u>:

    $\dfrac{40 \text{ seconds}}{60 \text{ sec/min}} = 0.67 \text{ min}$

    Now calculate flow rate from the faucet:

    $\text{Flow, gpm} = \dfrac{\text{Volume, gal}}{\text{Time, min}}$

    $= \dfrac{1 \text{ gal}}{0.67 \text{ min}}$

    $= 1.5 \text{ gpm}$

3. $\text{Normality} = \dfrac{\text{No. of Equivalents of Solute}}{\text{Liters of Solution}}$

    $= \dfrac{2.2 \text{ Equivalents}}{1.3 \text{ liters}}$

    $= 1.7 \, N$

4. The time must <u>all</u> be expressed in terms of minutes:

$$\frac{8 \text{ seconds}}{60 \text{ sec/min}} = 0.1 \text{ min}$$

Therefore the total time is 1.1 min. The flow rate can now be determined:

$$\text{Flow, gpm} = \frac{\text{Volume, gal}}{\text{Time, min}}$$

$$= \frac{1 \text{ gal}}{1.1 \text{ min}}$$

$$= 0.9 \text{ gpm}$$

5. $$\text{Estim. } Cl_2 \text{ Resid., mg/}L = \frac{(Cl_2 \text{ Resid., mg/}L)(\text{Distilled } H_2O, \text{ m}L)}{(\text{Sample Vol., drops})\left(\dfrac{0.05 \text{ m}L}{\text{drop}}\right)}$$

$$= \frac{(0.3 \text{ mg/}L)(10 \text{ m}L)}{(2 \text{ drops})(0.05 \text{ m}L/\text{drop})}$$

$$= 30 \text{ mg/}L$$

6. Before pipe volume and flushing time can be calculated, the diameter of the pipe must be expressed in terms of feet:

$$\frac{(0.75 \text{ inches})}{12 \text{ inches/ft}} = 0.06 \text{ ft}$$

Now flushing time can be calculated:

$$\text{Flushing Time, min} = \frac{(0.785)(D^2)(\text{Length, ft})(7.48 \text{ gal/cu ft})(2)}{\text{Flow Rate, gpm}}$$

$$= \frac{(0.785)(0.06 \text{ ft})(0.06 \text{ ft})(70 \text{ ft})(7.48 \text{ gal/cu ft})(2)}{0.5 \text{ gpm}}$$

$$= 5.9 \text{ min}$$

7. $$\frac{\text{Desired Wt., grams}}{\text{Desired Sol'n Vol., m}L} = \frac{\text{Actual Wt, grams}}{\text{Actual Sol'n Vol., m}L}$$

$$\frac{50 \text{ grams}}{1000 \text{ m}L} = \frac{44.19 \text{ grams}}{x \text{ m}L}$$

$$(x)(50) = (44.19)(1000)$$

$$x = \frac{(44.19)(1000)}{50}$$

$$x = 884 \text{ m}L$$

8. Use the same equation, fill in the given data, then solve for the unknown value.

$$\frac{\text{Flow,}}{\text{gpm}} = \frac{\text{Volume, gal}}{\text{Time, min}}$$

$$0.7 \text{ gpm} = \frac{1 \text{ gal}}{x \text{ min}}$$

$$(x)(0.7) = 1$$

$$x = \frac{1}{0.7}$$

$$x = 1.4 \text{ min}$$

9. First convert 750 mL to liters:

$$\frac{750 \text{ m}L}{1000 \text{ m}L/L} = 0.75 \, L$$

Then calculate the normality of the solution:

$$\text{Normality} = \frac{\text{No. of Equivalents of Solute}}{\text{Liters of Solution}}$$

$$= \frac{1.1 \text{ Equivalents}}{0.75 \text{ Liters}}$$

$$= 1.5 N$$

10. Estim. Cl$_2$ Resid., mg/L $= \dfrac{(\text{Cl}_2 \text{ Resid.,}) \quad (\text{Distilled H}_2\text{O,})}{(\text{Sample Vol.,}) \ (0.05 \text{ m}L)}$
$$\dfrac{\text{mg}/L \qquad\qquad \text{m}L}{\text{drops} \qquad\qquad \text{drop}}$$

$$= \dfrac{(0.4 \text{ mg}/L) \ (10 \text{ m}L)}{(3 \text{ drops}) \ (0.05 \text{ m}L/\text{drop})}$$

$$= 27 \text{ mg}/L$$

11. **Step 1:** (Add 40°)

$$\begin{array}{r} 72° \\ +\ 40° \\ \hline 112° \end{array}$$

Step 2: (Multiply by 5/9 or 9/5)

In this example the conversion is from Fahrenheit to Celsius. Since the answer should be a **smaller number**, multiply by 5/9:

$$\dfrac{(5)}{9} \ \dfrac{(112°)}{1} = \dfrac{560°}{9}$$

$$= 62°$$

Step 3: (Subtract 40°)

$$\begin{array}{r} 62° \\ -\ 40° \\ \hline 22° \end{array}$$

$$72° \text{ F} = 22° \text{ C}$$

12. The flushing time required is:

$$\text{Flushing Time, min} = \dfrac{(0.785) \ (D^2) \ (\text{Length, ft}) \ (7.48 \text{ gal/cu ft}) \ (2)}{\text{Flow Rate, gpm}}$$

$$= \dfrac{(0.785) \ (0.06 \text{ ft}) \ (0.06 \text{ ft}) \ (50 \text{ ft}) \ (7.48 \text{ gal/cu ft}) \ (2)}{0.5 \text{ gpm}}$$

$$= 4.2 \text{ minutes}$$

Convert the fractional part of a minute (0.2 min) to seconds:

$$(0.2 \text{ min}) \ (60 \text{ sec/min}) = 12 \text{ sec.}$$

Therefore the total flushing time required is:

$$4 \text{ min } 12 \text{ sec}$$

13.

$$\frac{\text{Desired Wt., grams}}{\text{Desired Sol'n Vol., m}L} = \frac{\text{Actual Wt, grams}}{\text{Actual Sol'n Vol., m}L}$$

$$\frac{3.27 \text{ grams}}{1000 \text{ m}L} = \frac{2.96 \text{ grams}}{x \text{ m}L}$$

$$(3.27)(x) = (2.96)(1000)$$

$$x = \frac{(2.96)(1000)}{3.27}$$

$$x = 905 \text{ m}L$$

14. Set the normality and volume of the first solution equal to the normality and volume of the second solution:

$$N_A V_A = N_B V_B$$

$$(0.2)(x \text{ m}L) = (0.01)(400 \text{ m}L)$$

$$x = \frac{(0.01)(400)}{0.2}$$

$$= 20 \text{ m}L \text{ NaOH}$$

15. $\text{Estim. Cl}_2 \text{ Resid.,} = \dfrac{(\text{Cl}_2 \text{ Resid.,})(\text{Distilled H}_2\text{O, })}{\underset{\text{drops}}{(\text{Sample Vol.,})} \underset{\text{drop}}{(0.05 \text{ m}L)}}$
$\qquad\quad \text{mg}/L \qquad\qquad\qquad \text{mg}/L \qquad\qquad\quad \text{m}L$

$$= \frac{(0.4 \text{ mg}/L)(10 \text{ m}L)}{(4 \text{ drops})(0.05 \text{ m}L/\text{drop})}$$

$$= 20 \text{ mg}/L$$

16.

$$\frac{\text{Desired Wt., grams}}{\text{Desired Sol'n Vol., m}L} = \frac{\text{Actual Wt, grams}}{\text{Actual Sol'n Vol., m}L}$$

$$\frac{7.6992 \text{ g}}{1000 \text{ m}L} = \frac{7.2725 \text{ g}}{x \text{ m}L}$$

$$(x)(7.6992) = (7.2725)(1000)$$

$$x = \frac{(7.2725)(1000)}{7.6992}$$

$$x = 945 \text{ m}L$$

17.   **Step 1:** (Add 40°)

$$\begin{array}{r} 18° \\ +\,40° \\ \hline 58° \end{array}$$

Step 2: (Multiply by 5/9 or 9/5)

In this example the conversion is from Celsius to Fahrenheit. Since the answer should be a **larger number**, multiply by 9/5:

$$\frac{(9)}{5}\ \frac{(58°)}{1} = \frac{522°}{5}$$
$$= 104°$$

Step 3: (Subtract 40°)

$$\begin{array}{r} 104° \\ -\ 40° \\ \hline 64° \end{array}$$

Therefore 18° C = 64° F

18.   **Step 1:** (Add 40°)

$$\begin{array}{r} 80° \\ +\,40° \\ \hline 120° \end{array}$$

Step 2: (Multiply by 5/9 or 9/5)

In this example the conversion is from Fahrenheit to Celsius. Since the answer should be a **smaller number**, multiply by 5/9:

$$\frac{(5)}{9}\ \frac{(120°)}{1} = \frac{600°}{9}$$
$$= 67°$$

Step 3: (Subtract 40°)

$$\begin{array}{r} 67° \\ -\ 40° \\ \hline 27° \end{array}$$

Therefore 80° F = 27° C

Appendices

APPENDIX 1

Column Friction Loss (in Ft) per 100 Ft of Column

Column Size	4"			5"			6"			8"			10"				12"				14" OD			
Tube Size	1¼"	1½"	2"	1¼"	1½"	2"	1½"	2"	2½"	2"	2½"	3"	2"	2½"	3"	3½"	2"	2½"	3"	3½"	2½"	3"	3½"	4"
50[b]	.65	.86	1.6																					
75	1.3	1.7	3.3																					
100	2.2	2.8	5.3	.54	.65	.94																		
125	3.2	4.2	7.8	.81	.96	1.4																		
150	4.4	5.8		1.1	1.3	1.9																		
175	5.8	7.5		1.5	1.7	2.5																		
200	7.3	9.4		1.8	2.2	3.1	.73	.96	1.4															
225				2.3	2.7	3.9	.90	1.2	1.7															
250				2.7	3.3	4.7	1.1	1.4	2.0															
275				3.3	3.9	5.6	1.3	1.7	2.4															
300				3.8	4.5	6.4	1.5	2.0	2.8															
325				4.4	5.2	7.4	1.7	2.3	3.2															
350				5.0	6.0	8.4	2.0	2.6	3.6															
375				5.6	6.7	9.5	2.2	2.9	4.1															
400				6.3	7.5		2.5	3.3	4.6	.61	.74	1.0												
450				7.8	9.3		3.1	4.1	5.7	.77	.91	1.3												
500							3.7	5.0	6.9	.93	1.1	1.5												
550							4.4	5.8		1.1	1.3	1.8												
600							5.2	6.8		1.3	1.5	2.1												
650							6.0			1.5	1.8	2.5												
700										1.7	2.0	2.8												
750										1.9	2.3	3.2												
800										2.2	2.6	3.6	.57	.65	.77	.95								
850										2.4	2.9	4.0	.63	.72	.86	1.1								
900										2.7	3.2	4.5	.70	.80	.96	1.2								
950										2.9	3.5	4.9	.77	.88	1.1	1.3								
1000										3.2	3.9	5.4	.85	.97	1.2	1.4	.34	.38	.44	.50				
1200										4.5	5.4	7.6	1.2	1.4	1.6	2.0	.47	.54	.62	.71				
1400										6.0	7.2	10.	1.6	1.8	2.2	2.7	.62	.71	.82	.94				
1600										7.6	9.1	13.	2.0	2.3	2.8	3.4	.80	.90	1.1	1.2	.47	.53	.59	.67
1800										9.4	11.		2.5	2.8	3.4	4.3	.99	1.1	1.3	1.5	.58	.65	.73	.84
2000										11.	13.		3.0	3.5	4.2	5.2	1.2	1.4	1.6	1.8	.71	.80	.89	1.0
2200													3.6	4.1	5.0	6.1	1.4	1.6	1.9	2.1	.85	.95	1.1	1.2
2400													4.2	4.9	5.8	7.2	1.7	1.9	2.2	2.5	.99	1.1	1.2	1.4
2600													4.9	5.6	6.8	8.2	1.9	2.2	2.5	2.9	1.1	1.3	1.4	1.6
2800													5.6	6.4	7.8	9.6	2.2	2.5	2.8	3.3	1.3	1.5	1.6	1.9
3000													6.4	7.4	8.8	10.	2.5	2.9	3.3	3.8	1.5	1.7	1.9	2.1
3200																	2.8	3.2	3.7	4.3	1.7	1.9	2.1	2.4
3400																	3.2	3.6	4.2	4.8	1.9	2.1	2.4	2.7
3600																	3.5	4.0	4.7	5.3	2.1	2.4	2.6	2.9
3800																	3.9	4.4	5.1	5.9	2.3	2.6	2.9	3.3
4000																	4.3	4.9	5.6	6.4	2.5	2.9	3.2	3.6
4200																	4.7	5.3	6.2	7.1	2.8	3.1	3.5	3.9
4400																	5.1	5.8	6.7	7.7	3.0	3.4	3.8	4.3
4600																	5.6	6.3	7.4	8.4	3.3	3.7	4.1	4.6
4800																	6.0	6.8	7.9	9.0	3.5	4.0	4.4	5.0
5000																					3.8	4.3	4.8	5.4
5200																					4.2	4.7	5.2	5.9
5500																					4.6	5.1	5.7	6.4
5750																					5.0	5.5	6.2	6.9
6000																					5.4	6.0	6.7	

Source: Peabody Floway, Fresno, CA

APPENDIX 2
Pump Performance Curve

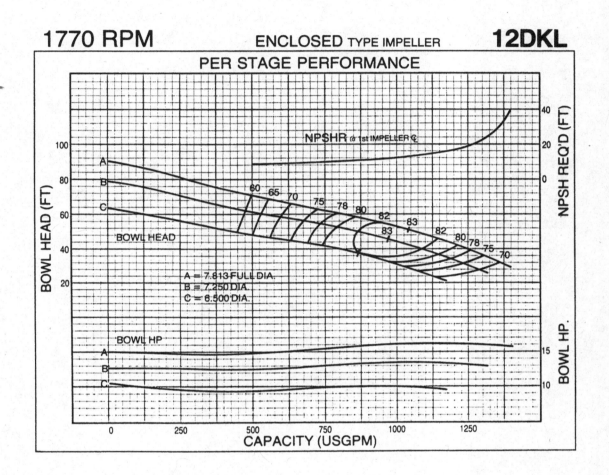

1770 RPM ENCLOSED TYPE IMPELLER **12DKL**

PER STAGE PERFORMANCE

APPENDIX 3-A
Fluoride Treatment Charts

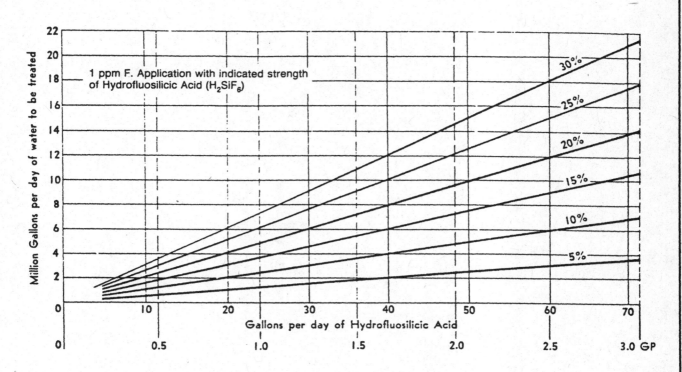

TREATMENT CHART I
Hydrofluosilicic Acid

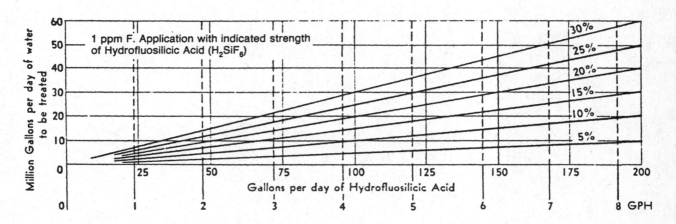

TREATMENT CHART II
Hydrofluosilicic Acid

Source: Wallace and Tiernan

APPENDIX 3-B
Fluoride Treatment Charts

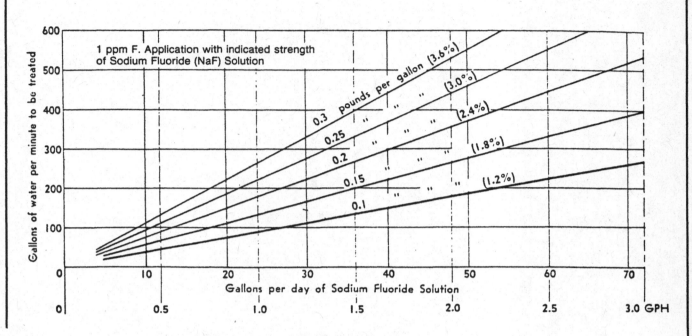

TREATMENT CHART III
Sodium Fluoride

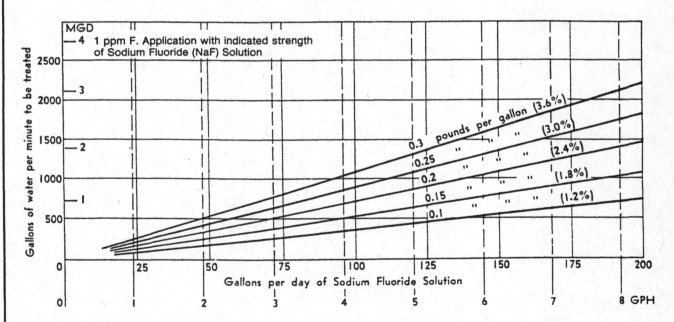

TREATMENT CHART IV

Source: Wallace and Tiernan

APPENDIX 4
Fluoridation Nomograph

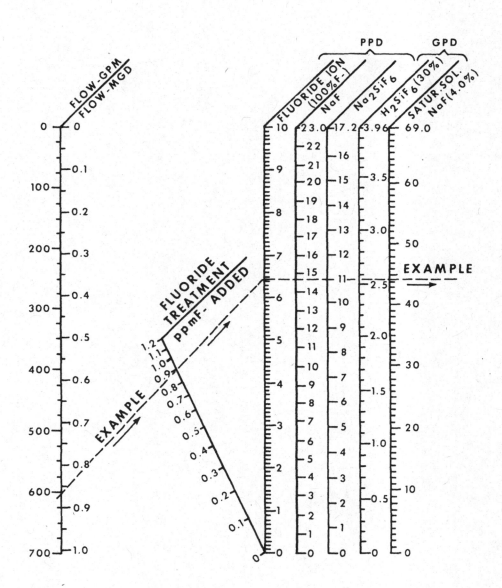

Source: *Fluoridation Engineering Manual*, EPA Publication EPA-520/9-74-022.

APPENDIX 5
Alkalinity Table

BICARBONATE, CARBONATE AND HYDROXIDE ALKALINITY

Alkalinity, mg/L as CaCO3			
Results of Titration	Bicarbonate Alkalinity	Carbonate Alkalinity	Hydroxide Alkalinity
P = O	T	O	O
P is less than 1/2 T	T – 2P	2P	O
P = 1/2 T	O	2P	O
P is greater then 1/2 T	O	2T – 2P	2P – T
P = T	O	O	T

Where   P = Phenolphthalein alkalinity
        T = Total alkalinity